GEOGRAPHY

History and Concepts

A Student's Guide

THIRD EDITION

ARILD HOLT-JENSEN

University of Bergen

English adaptation and translation
BRIAN FULLERTON
University of Newcastle upon Tyne

SAGE Publications

London · Thousand Oaks · New Delhi

Sage Publications Ltd
6 Bonhill Street
London EC2A 4PU

SAGE Publications Inc.
2455 Teller Road
Thousand Oaks, California 91320

SAGE Publications India Pvt Ltd
32, M-Block Market
Greater Kailash – I
New Delhi 110 048

British Library Cataloguing in Publication Data

A catalogue record for this book is available from the British Library

ISBN   0 7619 6179 8
ISBN   0 7619 6180 1 (pbk)

Library of Congress catalog record available

Typeset by Dorwyn Ltd, Rowlands Castle
Printed in Great Britain by The Cromwell Press, Trowbridge
A B C D E F   4 3 2 1 0 9

To my teacher in geography
Professor Fridtjov Isachsen
1906–79

Contents

Preface

TO THE STUDENT

Geography is intriguing because it is there all around you; the world that is an inextricable part of your everyday life. The fascination of the discipline is that it gives you keys which can help you to understand and explore that everyday world. To explain this fascination further I will briefly outline my own personal road to involvement with the discipline.

I started primary school in 1944 when Norway was closed off from the rest of the world as a result of German occupation. It was forbidden at that time to own a radio and to listen to the Norwegian broadcasts from London. My father, however, somehow managed to keep track of the changing war frontiers, which he marked with a thread of wool thumb-tacked to an old map of Europe. Then came May 1945, when Monty's young soldiers, who had actually taken part in the movement of that frontier thread, paraded our streets on National Day (May 17), wading through flowers thrown by we citizens who thronged the pavements to greet them. Russian prisoners held in my home town were set free and, before leaving, German soldiers befriended us children by distributing sweets. Europe happened once more around us and the world opened, as witnessed by the first American lorry (a 'Fargo') to arrive in our schoolyard and the docking of the first boat loaded with bananas. Then, in the summer of 1945, five volumes of *National Geographic Magazine* arrived which my father had subscribed to before the war. I had learnt to read Norwegian but not yet English. I could, however, 'read the maps' that were included in every second issue. Capital cities were marked with a star and these 'star cities' I learnt by heart! We were no longer geographically excluded; the world was once more ours to explore.

Maps of the Pacific Ocean were particularly intriguing in 1947. In that year Thor Heyerdahl and his crew (including a crewman from my home town) drifted on a balsa-wood fleet from Peru to French Polynesia. Heyerdahl's book *The Kon-Tiki Expedition* was the first 'grown-up' book I read. My road to geography, however, was for a while diverted to botany: my school herbarium acquired some really good additions as a result of a hiking tour I undertook with my mother into the Norwegian mountains. So my university studies started with botany and an MA in that discipline seemed inevitable. Only to complete my BA I would study geography. It was then I really struck lucky: until the year I started to study geography, geography in Norway had been divided in two separate disciplines – physical and cultural geography. In 1959, the year when I started, these were amalgamated into a so-called 'basic discipline' of unified geography. And thus we learnt to understand the landscape around us as the result of the combined forces of humanity and nature.

At first, physical geography was the easier to understand; cultural and human geography seemed much more vague and were more difficult to get to grips with. At Kolsås, which is close to Oslo, I got my first thrill of understanding. Here I realized that the layers of rocks I climbed over, created 600–300 million years ago and hence before the age of dinosaurs, still had an important bearing on the landscape of today. I learnt to interpret the geomorphologic features of the land-scape. Travelling became much more interesting. Having learnt something of oceanography and climatology, it became clear how the Humboldt Current could transport the Kon-Tiki fleet across the Pacific as well as provide the crew with all the fish they needed.

Geography had fired my imagination – even more so after Pierre Gourou and Torsten Hägerstrand had visited us as guest lecturers, and they were both human geographers. Gourou lectured in French, which I could not speak and could hardly understand. Even so, I can still recall his lecture on Belgium because of its gradual, systematic progression, documenting Belgium's problematic political geography and the multifaceted divisions between Vallonia and Flanders. Hägerstrand brought exciting new ideas on research methods, including the simulation of geographical diffusion employing advanced statistical tools. Silly me, why hadn't I studied statistics!

In 1962, midway through my work towards my MA thesis, I was lucky enough to obtain a grant to attend a student geographical conference in Saarland. During conference excursions we became acquainted with the German School of land-scape geography and hence learnt to use our eyes to interpret the cultural features of the landscape. (Manure heaps in front of barns meant active agriculture; barns converted to new uses meant commuting and 'social fallow'.) More importantly, we had to present our thesis work as a paper and in oral form. My rather tradi-tional regional geography approach elicited some very useful comments from an English student: 'Why don't you build up a hypothesis from the background features you explained in the introduction, and follow that as a thread all through?' This gave me the clue I was looking for. I found the hypothetic-deductive method very useful for my thesis: not because my general hypothesis of population decline was proven in the mountain farming district I was studying but rather because this marginal region also showed some population increase as a result of contingent, local circumstances that I had to explain.

When I started my career as a young geography lecturer in Bergen in 1965, the small staff of our geography department shared responsibilities for a staff seminar on Peter Haggett's book *Locational Analysis in Human Geography*. A whole new world of models and statistical techniques was opened up to us. At the same time I had to teach a wide variety of courses in 'basic geography' to the yearly intake of new students. As the only one in faculty with a natural science background in biology and with some knowledge of mathematics, I had to teach biogeography as well as cartography, geomorphology, oceanography and regional geography – all of which were outside my own specialism of applied human geography. But this broad teaching experience was very useful and an experience I deeply deplore that new staff do not receive these days. As the faculty acquired new specialist staff I, too, ceased to have such a broad-based involvement but the problem remains that strictly specialized research-based teaching tends to give students a rather fragmented basis from which to tackle the necessarily specialist research frontier. The aim of this book, therefore, is an attempt to keep this fragmented landscape of geography together in one piece. This is a book on geography, not on physical or human geography. It has tried to widen its scope beyond the Anglo-American world and it intends not only to cover the history of the discipline but also to explore current ideas and concepts. Because of this broad scope all specialists will, quite correctly, comment that this book does not cover their specialism properly. But this book was not written for them; it is an introductory textbook that has the intention of guiding students into the exciting world of geography.

The book also has one other basic intention that needs more lengthy explanation. We need to go back to 1968, the year when student uproar hit Paris and 'flower power' and the anti-Vietnam war movement hit San Francisco and other American cities. Bergen University had a whole week of 'critical discussions' led by the students. As a young lecturer who enthusiastically promoted the Haggettian approach to geography, this debate took me quite unprepared. At a meeting attended by other social science disciplines, I presented geography's spatial science achievements. The critical students dismissed me as hopeless, aiming their attacks instead at the well-known Norwegian political scientist, Stein Rokkan, who was also present. His models and his use of the hypothetic-deductive method were challenged fiercely, and I was shocked at this.

However, I learnt two things. The first was Rokkan's truly academic attitude. Even though the criticism was devastating he liked it. He liked the discussion and the criticism because it helped him to develop new ideas. He was a true supporter of Karl Popper's stand that an active and progressive science should be in a constant state of revolution. This is what distinguishes scientific discourse from political and other debates.

The other thing I learnt was that I had to have a better understanding of the positivism debate that had started earlier in sociology and political science but that was largely absent from geography. This led me to new reading, followed by lectures for my students and, eventually, to the first Norwegian edition of this book in 1976. The first English edition came, with considerable additions and alterations, in 1981. New directions and discussions within geography followed in the 1980s, which resulted in the second English edition in 1988 and the second Norwegian edition in 1990. In 1992 a Spanish edition even appeared, based on a translation from the English. In the eleven years since the second English edition was published, new geography students (as witnessed by continuing sales) have apparently found the book useful. On the other hand, I myself have increasingly found the 1988 edition unsatisfactory. To a large degree it has become outdated as the discipline continues its progression through a constant state of revolution. To some extent I also feel there were some misinterpretations in the text, which further reflection has helped me to correct. As all authors must feel, a book is never finished and complete; when a book is finally published, you have moved on, and have possibly learnt new things.

As an observer of academic development during the more than thirty years since 1968 it has, however, been sad to see so many of those young student critics of 1968 come to adhere to new orthodox trains of thought. They had not the stamina of Rokkan and Popper but built academic fortresses around their revolutionary gains, admitting into their fortresses only disciples of the 'true belief'. In the 1980s and 1990s poststructuralism and postmodernism have come to the fore in human geography. This, in principle, means the death of all orthodoxies and an openness to basic criticism, including self-criticism in all fields of inquiry. But it may also lead to scientific impotence: a situation in which we do not dare to continue an investigation, present its results or write it up as a thesis because we can see all its weaknesses and the potential for criticism from others. This frustrating situation must, of course, be avoided. But how can we proceed? This is the second basic intention of this book: a kind of solution, summarized in three points:

1) *Every scientific presentation is a simplification.* It is not the truth in itself, only a possible part-truth as seen through the disciplinary lens of the scientist. Simplification is necessary if we are to convey to an audience an understanding it can get to grips with. This is, of course, particularly the case with a textbook such as this: an introductory work has to simplify (and shorten) in order to be readable and comprehensive.

2) *You should not believe that something written in a textbook or presented by a senior academic is true simply because they say it is true.* You should only believe it when you have had the opportunity to test the postulates yourself. This may seem very difficult for a young student but it is, at the very least, a plea to keep an open mind and to remain sceptical to any form of orthodoxy. This is also a warning. When writing a thesis, it is too easy to hide behind references to leading iconoclasts or to hide in the fogs of obscure academic concepts. You must feel free to declare whatever findings and perspectives you can offer! It is hoped that you will be aided in this by those who adhere to the next point.

3) *Criticism should be positive and constructive rather than negative.* This, I think, is difficult for most academics when assessing and marking dissertations or student essays. It is easy to recognize failures, missing perspectives and misunderstandings and to focus one's assessment and criticism on these. Students find it equally easy to criticize a lecturer for delivering a lousy lecture that was 'so badly structured'. Assessment, however, has a human side too. You are stuck with that lecturer and no one performs better if he or she only hears negative criticism. When a teacher supervises student theses, this becomes very obvious. You have to find something positive to say so that the students will perform to the best of their abilities. Look for positive points first: did I learn something new from this chapter or lecture? Was a new perspective or approach presented? Was the structure logical and consistent? Sum up the positive points first; then consider any deficiences.

When reading this book you will realize that these points are as equally important when we assess the history of the discipline. One-sided, negative criticism often leads to a new orthodoxy because such criticism is defensive; it is not open to fruitful discourse in which one continues to learn new things. And this is the whole point: the world of everyday environments is ours, and geography should help us to explore it. I suggest you look twice at Robert Sack's diagram (Figure 5.2 in this book). To ordinary citizens living 'somewhere', the geographic environment is one. It consists of nature, social relations and meanings (i.e. your own personal 'roots' in a place). There is no division between human and physical geography. As researchers, however, we take on a certain paradigmatic perspective when learning in academia. We look through an academic lens from 'nowhere' to the 'somewheres' of geographic everyday life. The important point is that any lens only conveys a part, perhaps even a distortion, of geographic reality. This should warn us against any kind of academic self-righteousness: 'we alone know'. I wish you luck with your further exploration of the 'somewheres' in this world of ours. And I hope you proceed further than I have during a lifetime in geographic academia.

TO THE TEACHER

This third edition of *Geography: History and Concepts* has been completely revised and rewritten in order to present new developments in geographical thought and methods. Those who have used the second edition will find that the framework of the first chapters (particularly of Chapters 2 and 3) has remained unchanged but the text is rewritten.

Chapter 1 has been rewritten to give a broader, general presentation of the discussions contained within the discipline, its subdivisions and its general position among the sciences. The aim of this chapter is to arouse the interest of the student for further travel into the history of the discipline; in order to understand the juxtaposition of different views in this introduction. Chapter 2 (covering the origins of geography) has been updated to include more recent understandings of the ideas promulgated by the discipline's founders. Chapter 3 (on paradigms and revolutions) now includes a more extensive discussion of critical rationalism and a new evaluation of the spatial science school. The first part of Chapter 4 (covering positivism and its critics) retains the former structure but the last part of this chapter is new; it contains sections on humanism, structuration theory, realism, poststructuralism and postmodernism.

In the second edition, Chapter 5 comprised a very substantial discussion of systems analysis and functionalism in geography – themes that are not covered in this new edition. Chapter 5 now consists of a completely new presentation of the basic concepts and themes in geographical research. Such concepts as relative and relational space, place, region, territory, landscape and local identity are discussed and their use in recent research is exemplified. This chapter also discusses gender issues in geographic research, geographic exclusion and 'otherness' in relation to place, and territoriality and the research impact of GIS.

A new feature of this edition are the text boxes, which elucidate particular research examples and, in some cases, the life histories of leading geographers. They also clarify particular research concepts and methods.

This edition also contains a glossary, in which central concepts are defined and explained and that also functions as a subject index. The personality index contains, as in former editions, short biographical notes as far as it has been possible to find such information.

These new features emphasize the fact that this book is intended as a textbook for students that should be useful in the first year of study.

This edition is, like the former editions, a result of a collaboration between Brian Fullerton and myself. Brian was, however, unable to take responsibility for shaping the text of the new Chapter 5. But as a result of the very thorough text editing undertaken by David Kershaw, I feel the ideas contained within geographical thinking are conveyed here in a readable and clear style. The responsibility for any defects or false conclusions rests, however, wholly with the author. Citations from German texts are translated by the author.

For work on this new edition I must particularly acknowledge the debts I owe to Britt Dale and Kenneth Olwig, who read and commented upon most of the manuscript; to Tor Halfdan Aase, who provided me with material for two 'boxes' and who helped me with the section on semiotics; to Anne Lucas, who provided a box on the use of GIS; and to Anders Löfgren for providing material for the box on ontology, epistemology and methodology. I would also like to thank some of my students, and in particular Gert Myhren, who read and suggested improvements to Chapter 5. Comments and advice on the former editions have also benefited this one, and my thanks go to Ove Biilmann, Moshe Brawer, Olavi Granö, Jens Chr. Hansen, Torsten Hägerstrand, Jan Lundquist, Michael Morgan, Richard Morrill, Hans Skjervheim, Wolf Tietze, Aadel Brun Tschudi and Peter Weichhart.

Undoubtedly, though, my biggest debt is to my wife, Elisabeth, still enduring life with her husband through 35 years of geographic absentmindedness and workaholicism.

Bergen, Norway, December 1998
Arild Holt-Jensen

Acknowledgements

The author and the publishers would like to thank the following for permission to reproduce copyright material:

Walter de Gruyter, Publishers, for Figure 1.2, from Schmithusen, J., *Allgemeine Synergetik*, 1976; and Figure 5.10, adapted from Hard, G., *Die Geographie, eine wissenschaftstheoretische Einfuhrung*, 1973.

Blackwell Publishers Ltd for Figure 1.4, from Duncan, S.S. and Goodwin, M., *Uneven Development and the Local State*, 1988, p. 50; Figure 4.3, from Stoddart, D.R. (ed.), *Geography, Ideology and Social Concern*, 1981, p. 24; and Figure 5.1, from Johnston, R.J. (ed.) *On Human Geography*, 1986, p. 450.

Verlag Franz Deuticke, for Figures 1.5, 1.7 and 1.8, from Weichhart, P., *Geographie in Umbruch*, 1975, pp. 11, 99 and 104.

Harper & Row, for Figure 1.6, from Haggett, P., *Geography: A Modern Synthesis*, 1983.

Westermann Lernspielverlag, for Figure 2.1, from Leser, H., *Geographie, Das Geographische Seminar*, 1980, pp. 68–9.

Edward Arnold for Figure 3.2, from Harvey, D., *Explanation in Geography*, 1969, p. 34; and Figures 3.12 and 3.13 from Haggett, P., Cliff, A.D. and Frey, A., *Locational Analysis in Human Geography*, 2nd ed., 1977.

Random House UK Ltd, for Figure 3.3, from Popper, K.R., *The Logic of Scientific Discovery*, 1972, p. 243; and Figure 4.6, from Sayer, A., *Method in Social Science: A Realist Approach*, 1984, p. 215.

Studentlitteratur, Sweden, for Figure 3.5, from Alvesson, M. and Skoldberg, K., *Tolkning och reflektion. Vetenskapsfilosofi och kvalitativ metod*, 1994, p. 45.

Elsevier Science Ltd, for Figure 3.8, from Forer, P., *Progress in Human Geography*, Vol. 2, 1978, p. 247.

Henk Meijer, for Figure 3.11, from *IDG: Zuyder Zee/Lake Issjel*, 1981.

Geografforlaget, for Figure 3.14, from Reenberg, A., *Det Katastroferamte SAHEL*, 2nd ed., 1984, p. 15; and for Figure 2.1, from Biilmann, O. *Geografi, tradisjoner og perspectiver*, 1981.

University of Trondheim, for Figure 4.1, from Löfgren, A. *Om kvalitativ metod och fältarbete i geografi. Arbeider fra Geografisk Institutt, New Series C*, no. 4.

Prentice Hall Europe, for Figure 4.5, from Bhaskar, R., *A Realist Theory of Science*, 1975, p. 56.

Curzon Press, for Figure 4.9, from Manger, L. (ed.) *Local Islam*, 1999; and Figure 5.5, from Madsen, S. Toft (ed.), *State, Society, and the Environment*, 1999.

H. Meesenberg, for Figure 4.10, from *Geojournal*, 1996, Vol. 39, pp. 143–52.

Johns Hopkins University Press, for Figure 5.2, adapted from Sack, R.D., *Place, Modernity and the Consumer's World*, 1992.

Pearson Education Ltd, for Figure 5.3, from Porteous, J.D., *Environment and Behaviour; Planning and Everyday Urban Life*, 1977, p. 92; and Figure 5.8, from Dorling, D. and Fairbairn, D., *Mapping: Ways of Representing the World*, 1997.

Liverpool University Press, for Figure 5.6, from Herbert, D.T. and Raine, J.W., *Town Planning Review*, 1976, Vol. 47, pp. 325–38.

Methuen & Co. Ltd, for Figure 5.9, from Chorley, R.J., *Directions in Geography*, 1973, p. 38.

Swedish Society for Anthropology and Geography, for Figure 5.11, from Jones, M., *Geografiska Annaler*, 1988, Vol. 70B, pp. 197–204.

Every effort has been made to trace the copyright holders but if any have been inadvertently overlooked the publishers will be pleased to make the necessary arrangement at the first opportunity.

1

What is Geography?

Most people have only very vague notions about the content of scientific geography. School geography has left many with bad memories of learning the names of rivers and towns by rote. It is still common to meet people who think that geographers must have to learn a mass of facts, must know the population of towns all over the world and can name and locate all the new states in Africa. This idea of geography as an encyclopaedic knowledge of places is illustrated when a newspaper rings up its local department of geography to find out how many towns there are in the world called Newcastle, or when readers write in to settle bets as to which is the world's longest river.

People also have an idea that geography has something to do with maps. Less cynically than Swift:

> So geographers, in Afric-maps
> And o'er unhabitable downs
> With savage-pictures fill their gaps
> Place elephants for want of towns.

Geographers are thought to be people who know how to draw maps and are somehow associated with the Ordnance Survey or the US Coast and Geodetic Survey.

A third view is that geographers write travel descriptions – a reasonable belief for anyone who reads reviews of the year's books and sees that most of those listed under 'geography' are accounts of exciting expeditions to the Amazon, sailing trips around the world or something similar.

Each of these three popular opinions as to what geography is has some truth in it. The names and locations of towns are facts for geographers of the same order as dates are facts for historians. They are the basic building blocks of the subject, but they are not the subject itself. The map, which represents a collection of such data, is the geographer's most useful specific resource.

Different types of thematic maps are also important means of expression in geographical research, along with tables, diagrams and written accounts. The art of visual expression is much more closely associated with geography than with other social and natural sciences. Observations recorded during travel and field-work still provide essential data for geographers. A cultivation of the power of observation is therefore an important objective in the education of a geographer. Geographical training should develop the ability to 'see geographically', to observe and interpret a natural or cultural landscape in the field and/or through the study of maps, aerial photographs and other visual representations. Travel writers often lack this ability to observe and interpret. Admittedly, many people have had their initial interest in geography aroused through reading travel stories in their youth. But those who later trained as geographers eventually came to find this type of light reading unsatisfactory because of its lack of systematic observation.

EXPLORATION AND THE COSMOGRAPHIC TRADITION

Before this century, however, voyages of discovery and the mapping of formerly unknown lands *were* closely associated with geography. Wayne K. Davies (1972, p. 11), for instance, maintains that geography enjoyed its strongest relative position among the sciences during the so-called 'golden age' of exploration from the fifteenth to the nineteenth centuries. This was not due to the academic status of the subject during this period but to the work of a number of people who were actively involved with the mapping and description of the new lands being discovered. To the extent that they were working scientifically they would, however, be better described as **cosmographers** rather than geographers. **Cosmography**, as termed by Schmithüsen (1976, p. 10), included not only geography and cartography but also natural sciences like biology, geology and geophysics, and social sciences like anthropology, which only achieved their independent academic standing towards the end of the nineteenth century. Exploration, and all these other fields of cosmographic activity, were also regarded as being part of geography by the general public because they were carried out, to a large extent, under the auspices of the **geographical societies** (see Box 2.3 p. 29).

Geography developed as an academic discipline partly on the basis of a cosmographic philosophy that was developed to give coherence to the geographical societies' different activities. Gradually theoretical studies made an increasing contribution to the advancement of a specific geographical methodology. The chief emphasis remained, however, on geography as a **science of synthesis**, a science linking humanity and environment and creating a bridge between the social and natural sciences. Later parts of this book will show that **geographical synthesis** is not an easy task. Some even argue that it is impossible. We will, however, leave these critics for the time being, and present some of the features that characterize geography as a science of synthesis.

A SCIENCE OF SYNTHESIS

First, a student immediately notices that geography has no obvious place in the traditional classification of the sciences by faculty. Some parts of geography have their strongest affiliations with mathematics and natural sciences, others with history, philosophy and social sciences. Other sciences study distinctive types of phenomena: geologists study rocks, botanists plants, sociologists social groups and so on. The work of geographers involves several types of phenomena, each already studied by another science. Are geographers, therefore, 'jacks of all trades and masters of none'?

Some geographers argue that the subject-matter is shared with other disciplines but is treated in a different way for geographical purposes. Others affirm that the subject-matter of geography is exclusive; geographers alone study **places**.

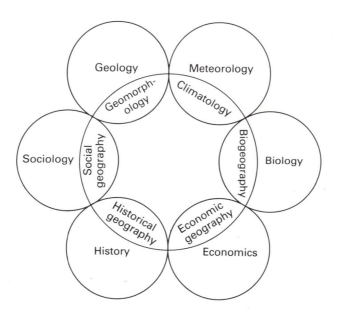

Figure 1.1 The circumference of geography
Source: Adapted from Fenneman, 1919

To clarify this point we may analyse the position of geography as seen by Fenneman (1919) (Figure 1.1). The diagram shows how sciences overlap and that each one of the overlapping zones, which also represent specialized, **systematic branches of geography**, belongs equally to some other science. The question is whether geography occupies anything more than these overlapping zones. Is anything left in the centre that is specifically geographical? What would be missing if the geography of an area were written by a group of scientists who each contributed a chapter: the geologist writing about the rocks, the botanist about the plant life, the meteorologist about climate, the demographer about population conditions and the economist about economic conditions? Such an account would surely fail to consider the overall interaction between phenomena. It is easy to see, for example, that the relationship between climate and soil type must have an important bearing on conditions of agricultural production and that the development of industry in an area may not be due only to economic factors but also to the natural resources of the area, its population potential and its historical and

Box 1.1   Categories of basic reasoning, or arguments for a chorologic, regional approach

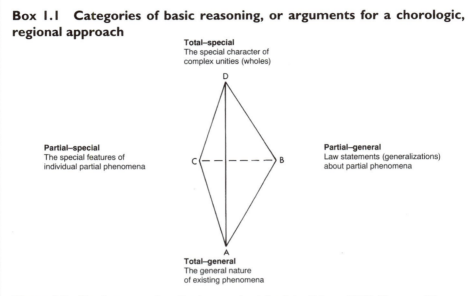

Total–special
The special character of
complex unities (wholes)

Partial–special
The special features of
individual partial phenomena

Partial–general
Law statements (generalizations)
about partial phenomena

Total–general
The general nature
of existing phenomena

Figure 1.2   The four categories of basic reasoning (after Schmithüsen, 1976). The use of the tetrahedron stresses that it is not necessary to choose only one of the four categories of basic reasoning. Most scientific reasoning involves more than one category

Joseph Schmithüsen, in his *Allgemeine Geosynergetik* (1976), attempts to establish a philosophical base for geography as a science, arguing that scientific research is based upon four different categories of basic reasoning, which are characterized by the following pairs of concepts: total–general, partial–general, partial–special and total–special (Figure 1.2). **Total-general reasoning** implies a holistic understanding of objects with the intention of making general statements about them. The questions asked are about the nature or essence of the objectives of science. This is the type of reasoning undertaken by the philosophers of science.

Partial-general reasoning implies a study of the parts as such, with the ultimate aim of presenting general statements, like laws in physics. Scientific reasoning in physics thus corresponds to B in Figure 1.2. **Partial-special reasoning** is concerned with the understanding of individual phenomena, such as 'why was Hadrian's Wall built across Britain?' This form of reasoning is also termed **idiographic**. **Total-special reasoning** tries to understand the complex features of particular unities, for example, landscapes or regions in geography.

According to Schmithüsen (*ibid.*, p. 22) the methodological peculiarity of geography is that it seeks to understand the complexities that exist together in a part of the earth in its spatial integration. This is in his opinion the basis of geography as an independent science. No other discipline concerns itself with the earth's surface and its spatial parts in their totality: with the association of phenomena in particular spaces or regions.

political development. Fenneman (*ibid.*) concluded that the interaction of all these factors can primarily be studied within definite areas or regions, and argued that geography should cultivate its core, **regional geography**, 'as a safeguard against absorption by other sciences'. Regional geography is defined as the study of areas in their total composition or complexity. In most cases, regional geography would, however, focus on the relationship between humanity and its habitat or another theme that makes an illuminating presentation of the region possible (Box 1.1; Figure 1.2).

While there is still some truth in the assessment that many geographers have lost their geographical identity to other subjects when working on the periphery, Fenneman's fear of absorption by other disciplines seems rather strange today. Instead of illustrating the robbers trying to disintegrate our discipline, Figure 1.1 may be used to demonstrate the point made by Ackerman (1958, p. 3) that geography is 'a mother discipline' from which other specialized disciplines like geodesy, meteorology, soil science, plant ecology and regional science have emerged. Geography has become an outward-looking discipline that has frequently created new specializations. This multidisciplinary perspective may be regarded both as our *raison d'être* and our life-raft in the sea of knowledge (Capelle, 1979, p. 65). If the periphery seems interesting, why not explore it; this will only widen the 'circle of geography'.

The systematic branches of human geography, especially, are breaking much new ground. Admittedly, useful knowledge from other disciplines can be fitted to the procedures used in geography but no other specific procedures are designed or followed in order to reveal the intricacies of spatial relations. Biology and economics have also treated spatial relations but in a much more limited way than geography has. Economic geographers are, for instance, concerned with structural changes in the **location** of industries and the **spatial factors** that may explain these changes, while economists are generally less interested in spatial factors and distributions, preferring to concentrate on the factors which determine economic development on the national scale.

There are even differences between botanists specializing in plant distribution and geographers interested in vegetation. In most cases, geographers will concentrate on vegetation types and their distribution, carrying their relevance to human geography somewhere at the back of their mind. Botanists, on the other hand, are more interested in the distribution of single plant species or grouping of species; vegetation as part of the landscape picture and its importance to humanity recede into the background; rare and inconspicuous flowers are often of greater interest to them than common and landscape-forming trees.

Geography exists to study variations in phenomena from place to place, and its value as an academic discipline depends on the extent to which it can clarify the spatial relations and processes that might explain the features of an area or a place. Geographical curiosity starts with the question, 'Why is it like this here?' Peter Haggett (1983) illustrates an approach to geographical inquiry by discussing the starting points of different scientists who might all be studying the same beach full of people bathing and sunning themselves (see Box 1.2). The geologist would be interested in the sand particles and the zoologist in the marine life along the shore. The sociologist might study the behaviour of the groups using the

Box 1.2   On the beach – a spatial science approach

Figure 1.3a   Regionalization of the beach

Figure 1.3b   Map of the beach; Each dot: the location of a person when the air photo was taken

Air photos make good starting points for geographers. A sequence of air photographs can show population densities at different times of the day. The study area can be reduced to a scale that is easy to work with later. While many scientists like to enlarge objects in order to study them, geographers often prefer to reduce the scale of complex phenomena through maps and photographs. Figure 1.3 shows a bathing beach on a fine summer's day. The natural landscape is a sandy beach with rocky promontories: fields and woodlands lie behind. We can also see a **cultural landscape** of many physical objects created by humans: roads, restaurants, a springboard and a pier, and we see people in the sea and sun-bathing, queuing up at the ice cream stalls and walking around.

The first reaction of geographers would presumably be to map accurately the exact location of each activity and map the changing population density. At the second stage, geographers would try to systematize what they observed on the photographs or maps into some sort of geographical pattern, for instance, by dividing the beach into zones of different population density. The third task would be to explain how the geographical pattern of density distributions came about. That explanation would consider a range of distinctive factors, including natural factors like exposure to sunshine and shelter provided by the topography, as well as cultural factors, such as the distance from restaurants, car parks and toilets. A crucial factor would be the **process of change**; each picture of the 'settlement pattern' on the beach would depend on what time of day the photo was taken.

The beach is, of course, a microcosm; geographers usually work with larger areas, analysing population densities by counties or the distribution of agricultural land across the world. However, studies of large areas are basically similar to those used to study the beach. The analysis of geographical patterns requires

1) an understanding of maps, projections and scales and of how maps are made;
2) knowledge of the statistical methods available to sort out those factors which might explain the patterns observed; and
3) an appreciation of the techniques which might elucidate the dynamics of changing geographical patterns.

Source: Based on Haggett, 1983.

beach and the economist might well be concerned with the marginal costs of the different ice-cream salesmen. For a geographer, one interesting field of inquiry would be to study the variations in population density of the different parts of the beach by mapping the location of each person. The geographer would find it difficult to work on the ground and might want a general oversight, perhaps using a helicopter to cover the situation by taking aerial photographs.

This example illustrates a working sequence in geographical research generally related to the **spatial science school** and the concept of **relative space** (see pp. 73–76): *localization → geographical pattern → explanation*. We start to locate spatial phenomena and use different mapping methods to discover specific geographical patterns, which we set out to explain. Because geographical patterns in general change through time, an understanding of the processes of change is crucial to the explanation of a specific geographical pattern.

This working sequence is still rewarding as an educational device, but it no longer serves as the prime model for research work; today a geographical research project often starts with the presentation of a social or natural process which seems to be related to geographical factors or have spatial relevance. Examples would be changing patterns of manufacturing or changing extensions of glaciers. Analysing and perhaps focusing on general trends of change, we may find that there are geographical differences in the patterns of change. While the ice caps of Antarctica and the North Pole are diminishing, the glaciers in Norway are at the same time increasing. While textile mills have been closed down all over Britain, there are still some pockets where the industry thrives. How can we explain such geographical differences? To what extent do local factors matter in global

Box 1.3   Pit and place: reflections on the geographical patterns of the miners' strike of 1984–5

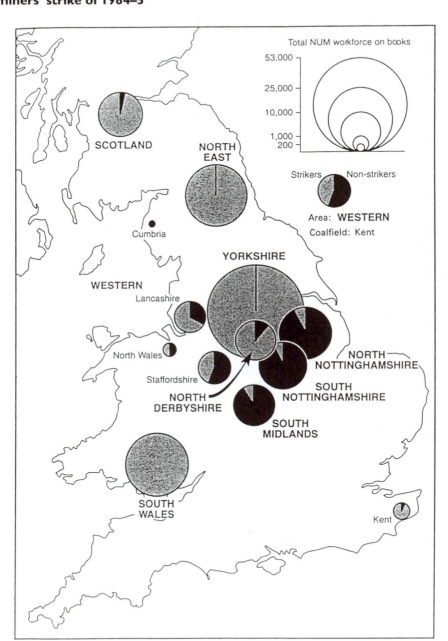

Figure 1.4   The distribution of support for the NUM strike in September 1984
Source: Duncan and Goodwin, 1988, p. 50; Johnston, 1991

In March 1984 the National Union of Mineworkers (NUM) called its workers to strike as a protest against the government plan for extensive pit closures. The strike lasted till March 1985, at which time the strikers went back to work, although no settlement was reached. The strike was a very important symbolic fight between the union and the Conservative government, a fact that should support full solidarity between the miners. In addition coal mines are generally located outside the main industrial conurbations with strong bonds of workplace, home and social life: a society very much based on solidarity. Emphasis on the crucial importance of the pit for the local community, its economic and social life, was the main argument in the strike leaders' campaign. What interests us here is that, in spite of this, there was an uneven geographical support for the strike. In some mining regions NUM got 100% support throughout the year-long dispute; in others the backing was small and coal continued to be mined. Figure 1.4 gives a picture of the geographical variability of the participation in the strike at one date – approximately halfway through its duration. Willingness to strike could to some extent be related to regional variations in the prospects of pit closures: South Yorkshire, north Nottinghamshire and north Derbyshire recorded surpluses and had good prospects for future working. But the 'self-interest' explanation fails as the strike had strong support in Yorkshire though not in Nottinghamshire. So the pattern must rather reflect long-established cultural differences between the mining districts. Johnston (1991, pp. 121–3) points out that the special system of subcontracting established in Nottinghamshire, at an early date, created a workforce divided into many small relatively independent units each with its own wage rate – creating no miner solidarity of the type developed in Yorkshire. An 'aftermath of the strike was the formation in Nottinghamshire of the Union of Democratic Miners (UDM), which was not federated with the NUM and was denied membership of both Trades Union Congress and the Labour party. Turning to the *politics of the community* the Dukeries [in north Nottinghamshire] again stands out as atypical among British coalfields' (*ibid.*, p. 129).

This example illustrates that local, geographically **contingent** determinants matter or, in the words of Peet (1998, p. 135), 'that the historical dynamics of socio-economic systems can be comprehended only in the geographical context because spatially specific circumstances limit the possibilities of human actions'. At the same time the example elucidates a presently more common approach in geographic research: from processes in space to geographical explanation.

processes of change? As exemplified in Box 1.3, we are here making use of the following work sequence: *process → variations of geographical change → explanation.*

All the phenomena which have a recognizable spatial distribution or can be shown on a map are basically of geographical interest, but many are of little significance. It is, however, impossible to say unambiguously which of the phenomena are of no significance. This has to be decided by the circumstances of each case. In most cases, for example, the distribution of Mormons in a region would have little geographical interest, but where, as in parts of the American West, their distribution becomes dominant or has contributed to the cultural and economic development of the area, then it becomes significant.

In research it is not possible to take account of all the factors of geographic interest. Geographers might restrict themselves to the changing distribution pattern of only one geographical phenomenon (for instance, the distribution of service, or shopping, centres in an area) and try to find explanations for it.

THE ORGANIZATIONAL PLAN OF GEOGRAPHY

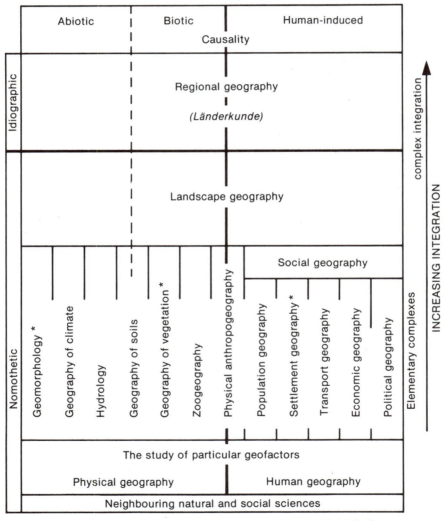

* Focusing themes in physical, eco- and human geography, respectively

Figure 1.5   An organizational plan of geography
Source: Uhlig, 1971; Weichhart, 1975

The number of branches of systematic geography could be argued indefinitely. In 1919 Fenneman even included 'mathematical geography', which has long been regarded as part of astronomy. Uhlig (1971, p. 15) limits the field of geographical inquiry to five aspects of the physical environment and five aspects of human life (Figure 1.5). Uhlig uses the term **social geography** in the German sense of integrative human geography rather than in the Anglo-American usage as a specialized branch of human geography. Of more significance is Uhlig's location of **landscape geography** at a higher level of integration than the systematic branches. In the German geographical tradition, landscape geography (*Landschaftskunde*) forms a transition or bridge between systematic geography and regional geography (*Länderkunde*) which is the most complex form of geographical integration. The landscape is seen as the product of the interaction of **geofactors** and only such features as are repetitive and appear in accordance with certain rules or laws are taken into account. The landscape concept represents an integration of human and physical geography that considers elements and properties (or characteristics) which form landscape types (Weichhart, 1975, p. 9). The aim of landscape geography is therefore to present and explain a typology of landscapes. Regional geography, on the other hand, seeks to give a total integrated picture of a specific area. Whereas landscape geography is concerned with the **nomothetic** (law-controlled) aspects of a certain area, the real research interest of regional geography lies in those aspects that make a region specific. Such **singular** aspects that have arisen in the course of history are the results of individual decisions through time and mould each region in a way peculiar to itself (*ibid.*, p. 16).

Many geographers have regarded regional geography as the core of the subject. Systematic geography becomes the area in which scientific laws are formulated and regional geography becomes the field in which such laws are tested, and which also provides a **synthesis** of the physical and human phenomena within an area or region (Hettner, 1927). It has, however, been difficult for regional geography to fulfil these roles in the field of research.

A NEW SYNTHESIS?

Haggett (1983) tried to develop a new form of synthesis which diverges from the traditional division of the subject (Figure 1.6). He emphasizes that the historical divisions are important if only because universities still often use them as a basis for their courses. But it is more valuable, he thinks, to divide the subject up in relation to the way in which it analyses its problems. His three main groups are defined as follows:

1) *Spatial analysis* concerns itself with the variations in the localization and distribution of a significant phenomenon or group of phenomena; for instance, the analysis of variations in population density or poverty in rural areas. Which factors control the distribution pattern? Which processes led to this pattern? How can these patterns be modified so that the distribution becomes more effective or just?

2) *Ecological analysis* concerns itself with the study of connections between human and environmental variables. In this type of analysis we are studying the

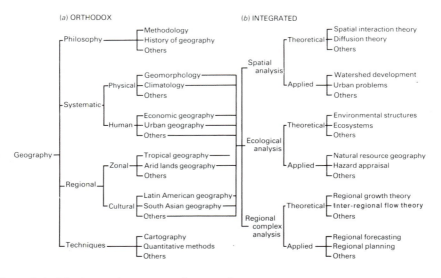

Figure 1.6 The internal structure of geography
Source: From Haggett, 1983

    relations within particular bounded geographical spaces, rather than the spatial variations between regions.

3) *Regional complex analysis* combines the results of spatial and ecological analysis. Appropriate regional units are identified by areal differentiations. Connecting lines and flows between the individual regions may then be observed.

In his textbook, Haggett (3rd edn, 1983) tried to arrange his themes under these headings. His work shows, however, that satisfactory models for complex regional analysis were not well developed. Weichhart (1975) attempted a better organizational plan for the discipline in order to promote geographical synthesis. He refutes the notion that 'geography as human ecology' or ecological analysis can form a leading concept for the whole of geography. Questions relating to humanity and its environment can only be a part of the research field – but need to become a more important part. Returning to Uhlig's (1971) organizational plan (Figure 1.5), Weichhart (1975) classifies the **geofactors** that are the important elements of the **geosphere** into three groups: abiotic, biotic and human induced. **Abiotic** factors are geology, soil, climatic features, ocean currents, etc.; **biotic** factors include vegetation, animal life and humans as biological creatures; and **human-induced** factors are settlement, transportation, industry, social structures, etc. Weichhart (*ibid.*, p. 98) maintains that it is possible on this basis to recognize four problem categories (Figure 1.7), under which most possible geographical research questions could be grouped. In **systematic geography**, at the lower stages of integration, we describe and explain, possibly through spatial theories or laws, the spatial variation of the **single geofactors**, be they abiotic, biotic or human induced. In this connection the other geofactors are taken into consideration only to the degree to which they influence the spatial variation of the geofactor under consideration. This contrasts with other categories of geographical problems where several geofactors are taken into account at the same time.

```
┌─────────────────────────────────────────────────────────────┐
│                    GEOGRAPHICAL QUESTIONS                     │
├─────────────────────────────────────────────────────────────┤
│  Complex relationships, system-related inter-relations and   │
│  interdependencies between one or more elements of:          │
│                                                               │
│  abiotic and biotic      abiotic, biotic and    human-induced │
│  factors                 human-induced          factors       │
│                          factors                              │
│                                                               │
│  Which generalizations                          Which         │
│  or laws?                                        generalizations│
│                                                  or laws?      │
│                          Which generalizations                │
│                          or laws?                             │
├─────────────────────────────────────────────────────────────┤
│  Chorological variations of individual elements of abiotic,  │
│              biotic and human-induced factors                │
│                                                               │
│            Which generalizations or laws?                    │
└─────────────────────────────────────────────────────────────┘
```

Figure 1.7   Problem categories within geography
Source: After Weichhart, 1975

Then we are looking for **complex system relations** that exist between several geofactors. Taking the division of labour within the discipline into consideration, Weichhart (*ibid.*, p. 99) envisaged three groups.

First it is possible to study the system relations between all or a number of the abiotic or biotic **geofactors** – geography as a physical geographical synthesis. The leading threads might include a **nomothetic**-orientated typology of natural landscapes or a process-orientated description of the evolution of the landscapes in a certain region.

On the human side, we may study the system relations between all or a number of the human-induced factors. The catalogue of relevant problems would, however, be incomplete without consideration of the inter-relations between abiotic, biotic and human-induced factors which constitute the human–environment system. Weichhart (*ibid.*, p. 98) makes the point that this does not imply that the totality of geofactors is brought into consideration. Geographical research has shown that the abiotic and biotic parameters needed to describe and explain the patterns of nature may not be identical to those needed to explain and describe the complex of relations between humanity and nature. Relief structures recede into the background; soil, vegetation and hydrological features become more important. Among human-induced factors, those which may be important to an understanding of social geographical structures may have less significance for the relations between humanity and nature. The study of humanity–environment relations is therefore not the same as a total synthesis of all the geofactors.

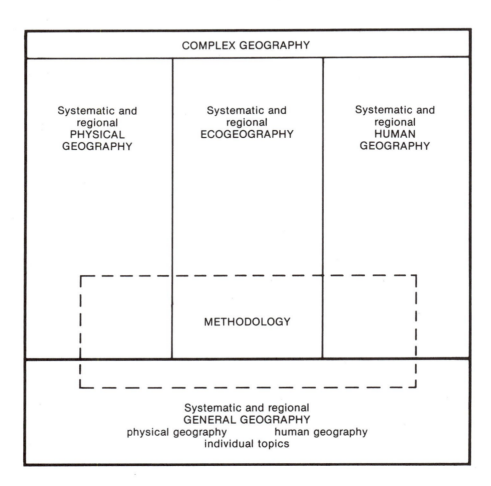

Figure 1.8   A modified organizational plan of geography
Source: After Weichhart, 1975

The original German organizational plan for geography shown in Figure 1.5 distinguishes between **general** or **systematic** (*allgemeine*) **geography**, and **special** or **regional geography**. Landscape geography is seen as a transition between the systematic and the regional. In real life, as Weichhart (*ibid.*, p. 100) observes, the general and the special are indissolubly joined.

For this reason, Weichhart (1975, p. 104), in his organizational plan for geography (Figure 1.8), made no distinctions between general and special geography but contrasted **allgemeine Geographie**, focusing on the study of a single geofactor in a regional or global context, with **complex geography**, which studies relations between several geofactors in a regional or world context. Complex geography is divided into three parts corresponding to the grouping of problems in Figure 1.7 and does not therefore represent a total synthesis of all possible geofactors but only of those of importance for the chosen problem field. The field of **ecogeography** (which Weichhart believes is becoming increasingly important) studies the relations between those geofactors of particular importance for humanity–environment problems (Box 1.4; Figure 1.9).

Box 1.4   Simplified examples of analysis models of geographical synthesis within physical geography, ecogeography and human geography

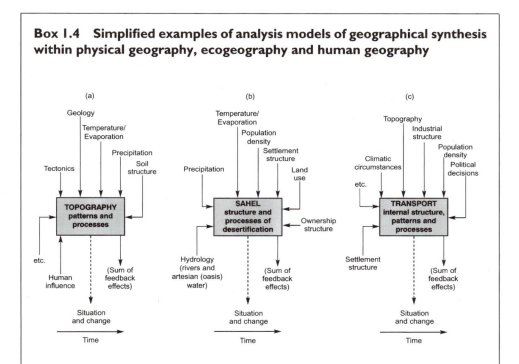

Figure 1.9   Simplified examples of analysis models of geographical synthesis within physical geography, ecogeography and human geography
Source: Leser, 1980, pp.68–9

Figure 1.9 illustrates some of the factors that have to be taken into account in the geographical study of the three concrete research themes:

a) The study of the topography of a particular area in physical geography.
b) The study of desert structure and desertification processes (for instance, in Sahel) in ecogeography.
c) The study of transport patterns and processes in human geography.

The figure should speak for itself; the main point here is that in all three fields of research both human and physical geographic factors have to be considered, but to different degrees. In ecogeography the integrative task is crucial, but we should note this does not mean we will need to integrate *all* the human and physical factors that are included in purer physical and human geographical research.

While Weichhart's organizational plan (Figure 1.8) may not be the ultimate answer to the problem of geographical synthesis, it provides a useful framework for the applied research projects that are of growing importance in geography. Many German departments use the threefold division of complex geography to provide a framework for those degree courses in geography that train planners for public and private service.

Recent decades have been marked by a general debate on the content and methods of geography in which the traditional subdivisions have been strongly criticized, and new theoretical approaches have been advanced. Before considering these ideas we will, in the next chapter, look at the historical traditions of geography.

2

The Foundation of Geography

GEOGRAPHY IN THE ANCIENT WORLD

Interest in geographical problems, and writings on subjects we can recognize as geographical, began long before the introduction of the subject into universities. In fact it is difficult to imagine how there were ever people who did not think geographically, who never considered the conditions under which they lived, and never wondered how people lived in other places. In this sense geographical thinking is older than the term **geography** (literally meaning 'earth description' from the Greek *geo*: earth and *grafein*: to draw), which was first used by scholars at the Museum in Alexandria about 300 BC.

Herodotus
The ancient Greeks made the first major contribution to the development of geography. Scholarly writers produced **topographical descriptions** of places in the known world, discussing both natural conditions and the culture and way of life of the people who lived there. The best known of these was Herodotus (*c.* 485–425 BC). He was first and foremost a historian, and is indeed often regarded as the father of history, but he placed historical events in a geographical setting; some of his writings are truly geographical in character. He not only described geographical phenomena as, for example, the annual flow of the Nile but also attempted to explain them.

Herodotus had no interest in those mathematical and astronomical problems that later became associated with geography – the measurement of the circumference of the earth and the establishment of exact locations for places. He accepted the Homeric view of the earth as a flat disc over which the sun travelled in an arc from east to west. The belief that the earth was a sphere was discussed in Herodotus' time but it was Aristotle, about a century after Herodotus, who produced observational evidence for a spherical form of the earth. The scholars at the Museum in Alexandria were then able to establish the foundations for the calculation of latitude, longitude and the size of the earth.

Eratosthenes
Eratosthenes (276–194 BC), who was chief librarian at the Museum, succeeded in calculating the circumference of the earth with remarkable precision. Of equal importance was his development of systems of co-ordinates for the world,

Box 2.1   How Eratosthenes calculated the size of the earth

Eratosthenes learned that the rays of the mid-day sun penetrated to the bottom of a well at Syene (Assuan) on the Tropic of Cancer, at the summer solstice. At the same time and on the same day he observed that the obelisk in front of the Museum in Alexandria cast a shadow. He used a plumb line to measure the angle of the sun's rays at that time and day, which turned out to be 1/50th of a circle. The north-south distance from Syene to Alexandria must be 1/50th of the circumference of the earth. The story goes that he sent some reliable slaves to Syene, travelling on an ox-cart with a lump on each wheel. They counted the jolts from the lumps to arrive at the number of wheel turns and, knowing the size of the wheel, calculated the distance from Alexandria to Syene. From this measurement, Eratosthenes calculated the circumference of the earth to be 252,000 stadii (a stadium equals 157.5 m) or c. 39,700 km. Modern measurements show the actual circumference to be 40,000 km. If you look closer at a map of Egypt you will find that Alexandria is located 7.8° north of the Tropic of Cancer, 1/50 of the circle or 7.2° is thus inaccurate. But this might have been counterbalanced by the fact that the road along the Nile does not go straight north–south.

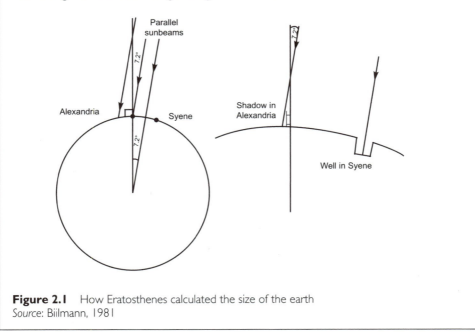

Figure 2.1   How Eratosthenes calculated the size of the earth
Source: Biilmann, 1981

i.e. latitude and longitude, which he used to locate places and to measure distances. This made it possible for him to draw the first passably accurate map.

Ptolemy

Eratosthenes' cartographical work was later developed by his students and successors at the Museum in Alexandria. Ptolemy (AD 90–168) wrote a major work in eight volumes which is now known as Ptolemy's '**Geography**'. The first volume explained the principles for calculating the dimensions of the earth, its division into degrees, calculations of latitude and longitude, and a discussion of map projections. The eighth volume contained maps of different parts of the world. Other volumes included tables of latitude and longitude for 4,000 places.

Although it was widely known how to calculate geographical latitude from the altitude of the sun in Ptolemy's time, calculations had only been made for a handful of places. The calculation of longitude was then only possible by estimating the length of journeys from one place to another, and so many of Ptolemy's locations for places were erroneous. His biggest mistake was to underestimate the size of the earth, rejecting the almost correct estimate of Eratosthenes in favour of a reckoning made by Posidonius in about 100 BC. Posidonius reckoned the circumference of the earth to be 180,000 stadii against Eratosthenes' estimates of 252,000 stadii. Because calculations of longitude had to be based on rather unreliable travel distances, Ptolemy's otherwise remarkably accurate map of the known world included too many degrees of longitude. The map extended from a prime meridian in the Canary Islands to a 180° meridian which crossed inner China. The actual distance between these places is only 120°.

Strabo

The topographical tradition of Greek geography, represented by Herodotus, was carried forward into Roman times. Strabo (64 BC–AD 20) wrote a work of 17 volumes called *Geographica*. This was largely an encyclopaedic description of the known world whose chief value was that it preserved for posterity many writings which he annotated and cited. *Geographica* also included attempts to explain cultural distinctiveness, types of governments and customs in particular places. The significance of natural conditions for cultural development was discussed in relation to a number of places, especially in the description of Italy.

Greek and Roman geographers concerned themselves with three basic issues:

1) The detailed **topographical description** of places and their history, which Ptolemy called **chorography**.
2) The measurement of the earth and the production of maps.
3) A more philosophical interest in the relations between humanity and the environment, which involved the beliefs that the earth demonstrates an order and purpose which has been designed by a deity, that the environment influences people and that people can only to some extent modify their environment.

These ideas have had an important bearing on the later development of geography.

MIDDLE AGES AND THE RENAISSANCE

The Middle Ages were a dark period for the development of science in Europe. At best, scholars made accurate but sterile copies of the works of the ancients, rejecting anything that did not conform with the dogmas of the Christian Church. Such an intellectual environment stifled any development of scientific investigation.

Concepts of the world which had been developed in ancient times were reshaped to conform to the teaching of the Church. The earth became a flat disc with Jerusalem at its centre.

Arab geographers
Ancient learning was, however, carefully preserved in the Islamic countries at the Arab universities of Spain, north Africa and the Middle East. Arab traders travelled widely and gathered information which could be used by scholars to fill in the gaps on Ptolemy's original map. The best known of these travellers was Ibn Batuta (1304–68), who travelled as far east as Peking in China, and sailed far south of the equator along the east coast of Africa. This particular trip showed that Aristotle had been wrong in believing that it was too hot here for human habitation, in what the Greeks had called the 'torrid zone'.

But already by the twelfth century, Al Idrisi (1099–1180) had shown that the Greek division of the world into five climatic zones (two cold, two temperate and one torrid) did not correspond to reality, and had suggested a more sophisticated world climatic system. He wrote a descriptive geography to accompany a large silver map of the world that had been constructed for Roger II of Sicily. His connection with this mighty Christian king made it possible for him to give an account of Europe in the first half of the twelfth century that is unrivalled (Unwin, 1992, p. 57).

Al Muqaddasi (945–88) was the most eminent Arab geographer of the Middle Ages, according to Scholten (1980). Before his time, geographers had generally based their topographical descriptions on compilations of second-hand sources. Muqaddasi was a pioneer of **fieldwork** and stated in his book *The Best Divisions of the Study of Climate* that he would not present anything as fact to his readers unless he had seen it with his own eyes. His book was based on extensive travels within the Islamic world.

The last great Islamic geographer of the Middle Ages was Ibn Khaldun (1332–1406). He established the foundation for historical geography in those of his writings which analysed the rise and fall of empires. He suggested that warlike nomads often founded large states, but after a while the nomads were absorbed by their permanently settled subjects. As peasants and townspeople, the rulers lost their warlike spirit and eventually their kingdoms fell apart. Ibn Khaldun both predicted and lived to see the collapse of the Islamic state he lived in. At the fall of Damascus in 1400 he actually met Tamerlane, the conqueror and devastator. It was unfortunate that the works of the Arab scholars were not translated into Latin or any other language until the nineteenth century; as a result, scholars in other parts of the world were unable to make use of their observations.

Europeans
The journeys undertaken by Europeans during the Middle Ages made little significant contribution to the development of geographical knowledge. Around AD 1000 the Norsemen sailed across the North Atlantic to Greenland and North America but the sagas of these adventures were only passed on by word of mouth and written down long afterwards in isolated Iceland. Similarly, the Irish sailors who crossed the Atlantic during the same period made little impact on geographical understanding about the world.

China

Meanwhile, exploration and learning flourished in China. Actually, Europe and India were 'discovered' by Chinese travellers long before Europeans reached the Orient. In the period between the second century BC and the fifteenth century AD, Chinese culture was the most efficient in the world in applying knowledge of nature to useful purposes. The study of geography was advanced well beyond anything known in Europe at this time; among other things, the Chinese used co-ordinates and triangulation to produce beautiful maps of China and neighbouring countries. When, however, the Italian Marco Polo (1254–1322) wrote an account of his travels to China describing the high level of Chinese learning, his book was widely discounted as a fictitious adventure.

The Renaissance

The Renaissance brought about a renewed interest in the geographical knowledge of ancient times. A copy of Ptolemy's *Geography*, which had been preserved in a monastery in Byzantium (Istanbul), was discovered and brought to Italy where it was translated into Latin in 1406. After the invention of printing in the 1450s, copies of this book, including reconstructed maps, were printed in Bologna, Rome and Ulm, and made a great impact on contemporary scholars (Bagrow, 1945). Columbus and other expeditionary leaders relied on Ptolemy's calculations (see above), but some of their more famous discoveries disproved the latter's calculations and changed the world picture he had established. There were new developments in cartography – new projections, especially that of Mercator in 1569, were invented; the first globes were made; and new world maps were published. The rediscovery of classical cartography played a significant role in the European expansion of political and economic power in the fifteenth and sixteenth centuries.

Few European explorers or scholars regarded themselves as geographers. They preferred the term **cosmography** (the descriptive science of the globe and its relation to the universe) which was used in a number of treatises, often modelled on the works of Strabo and Ptolemy.

The Renaissance saw a revival of another branch of ancient geography, that of **topographic description**. Accounts of voyages provided raw material for encyclopaedic works on the world, or on parts of it, often called **chorographies**. The jumble of information related to place-names included in such chorographies (where notes on natural and physical conditions were juxtaposed beside miscellaneous information on folklore, etymology and history) places them outside the field of science as we understand it today. The popularity of these works in their own time might well puzzle us until we realize that we

> are often prisoners of contemporary logic and cannot see those qualities of old works which cannot be integrated into our system . . . at the time it had just as much relevance as what continues to interest us today: it was part of what the episteme of the time indicated was knowledge

> (Claval, 1980, p. 379)

Strabo (64 BC–AD 20) had made a rough definition of the study of geography when he said that

> wide learning, which alone makes it possible to undertake a work on geography, is possessed solely by the man who has investigated things both human and divine . . . The

utility of geography . . . regards knowledge both of the heavens and of things on land and sea, animals, plants, fruits, and everything else to be seen in various regions – the utility of geography, I say, presupposes in the geographer the same philosopher, the man who busies himself with the investigation of the art of life, that is, of happiness.

(Cited in Fischer *et al.*, 1969, p. 16)

Above all geography retained in the fifteenth and sixteenth centuries the practical importance attributed to it by Strabo. It was essential for statesmen and merchants working towards the European domination of the globe. But as a science it was diffuse; there was no need to define disciplinary boundaries. Later disciplinary discussions on the content of the discipline have, however, been influenced by this early development; geography could be defined as the all-embracing **cosmography**, the science of the globe, or it could be **chorography**, the descriptive science of single regions (Box 2.2).

Box 2.2   *-GRAPHY or -LOGY?*

Well into the seventeenth century, geographical writing was profoundly influenced by Greek and Latin classical writers, and a number of classical Greek words were used to define aspects of the subject. Thus 'geography' combines *geo* (the earth) with *graphos* ('drawing' and, by extension, 'description'). 'Cosmography' involved the description of the universe (*cosmos*) and included the modern sciences of biology, cartography, geography, geophysics and also anthropology. *Chorography* (*choros* = place) described places in general and larger areas, whereas 'topography' was concerned with particular places at a smaller scale.

Words ending in *-logy* incorporate the Greek word *logos*, which signifies a rational principle or structure. (*Logos* is the 'Word' which was 'from the beginning' in the first chapter of St John's Gospel.) So *logos* implies explanation rather than the description implied by *graphos*. 'Chorology' suggests some explanation about places, just as 'chronology' deals with time and how historical events can be seen as logical outcomes of a course of previous events.

VARENIUS

Strabo (in the citation above) considered that geographers should study both human activities and natural conditions, a reasonable conclusion so long as geography was regarded as a descriptive discipline. The role of geography was to describe the earth and its different places, and in such descriptions both humanity and nature would play a role. After the Renaissance, as the natural sciences developed analytical methods, some scholars began to argue that the respective **methodologies** of physical and human geography should be different. In studies of natural phenomena, including climate, geology and landforms, it is possible to use the methods of the natural sciences and to draw conclusions with a large measure of scientific precision. The methods of natural science, however, do not lend themselves very well to the study of social and cultural phenomena. Our generalizations about human groups must be limited in time and space, and must relate to statements of probability rather than certainty.

Bernhard Varenius (1622–50), whose *Geographia Generalis* was published in Amsterdam in 1650, was one of the first scholars to suggest these essential differences in the character of physical and human geography. His book includes much material which today would be treated as mathematical geography or as astronomy. The *Geographia Generalis* is divided into three parts:

1) The absolute or terrestrial section, which describes the shape and size of the earth and the physical geography of continents, seas and the atmosphere.
2) The relative or cosmic section, which treats the relations between the earth and other heavenly bodies, especially the sun and its influence on world climate.
3) The comparative section, which discusses the location of different places in relation to each other and the principles of navigation.

Varenius intended that the *Geographia Generalis* should be followed by what he called, in his Foreword to the book, 'special' geography in which descriptions of particular places should be based upon

1) 'celestial conditions', including climates and climatic zones;
2) terrestrial conditions with descriptions of relief, vegetation and animal life; and also
3) human conditions including trade, settlement and forms of government in each country.

In fact he showed little enthusiasm to embark on the study of human geography as it was not possible to treat it in an exact way. He explained that he had included this last group of conditions as a concession to traditional approaches to the subject. Unfortunately, he did not have the opportunity to write on special geography since he died at the early age of 28 (Lange, 1961).

Varenius made two major contributions to the development of geography. First, he brought together contemporary knowledge of astronomy and cartography and subjected the different theories of his day to sound critical analysis. Secondly, his partition of geography into *general* and *special* sections originated what we now call systematic and regional geography. His *Geographia Generalis* dealt with the whole world as a unit, but was restricted to physical conditions which could be understood through natural laws. **Special geography** was primarily intended as a description of individual countries and world regions. It was difficult to establish 'laws' in the special (regional) part, for the explanation must be descriptive where people are involved. Preston James (1972, p. 124) points out that the general (systematic) and special (regional) parts of Varenius' work complement each other, and that Varenius saw general and special geography as two mutually interdependent parts of a whole.

THE PHILOSOPHER, IMMANUEL KANT

The lectures on physical geography given by Immanuel Kant (1724–1804) in Königsberg were of great significance for the later development of geography. The course, which was given more than 40 times, was one of his most popular. It was not a course on physical geography as we now understand it, because it also discussed human racial groups, their physical activities on the earth and natural conditions in the widest sense of the term. Kant regarded geography as only an approach to the empirical knowledge that was necessary for his philosophical research:

> But, finding the subject inadequately developed and organised, he devoted a great deal of attention to the assembly and organisation of materials from a wide variety of sources, and also to the consideration of a number of specific problems – for example, the detection of wind direction resulting from the Earth's rotation.
>
> (Hartshorne, 1939, p. 38)

Kant provided a philosophical foundation for the belief that the subject has a significant scientific contribution to make. He pointed out that there are two different ways of grouping or classifying empirical phenomena for the purpose of studying them: either in accordance with their nature, or in accordance with their position in time and place. The former is a **logical classification**, the latter a physical one. Logical classification lays the foundations for the systematic sciences – the study of animals is zoology, that of the rocks is geology, that of social groups is sociology and so on. **Physical classification** gives the scientific basis for history and geography: history studies the phenomena that follow one after the other in time (**chronological science**), while geography studies all phenomena belonging to the same place (**chorographic science**). This enabled him to argue that as empirical sciences 'geography and history fill up the total span of knowledge; geography namely that of space, but history that of time' (Kant, in May, 1970, p. 261).

Geographers have often appealed to Kant in justifying the existence of the subject and its special character amongst the sciences. Alfred Hettner (1859–1941), who did most to affirm the significance of Kant's work to geography, admitted (1927, p. 115) that for a long time he had not paid sufficient attention to Kant's exposition of geography but had later rejoiced to discover a close correspondence between the conclusions of the great philosopher and his own. The American geographer Richard Hartshorne closely followed Hettner in his *The Nature of Geography* (1939), which considers that Kant's approach led towards a satisfactory understanding of the nature of geography and answered all its basic questions.

Werlen (1993, pp. 2–3) maintains that Hettner misinterpreted Kant: 'For Kant geography was a descriptive or taxonomic discipline, rather than a science: it only had the status of a propaedeutic discipline. Kant used the word "chorographic", meaning descriptive, to describe geography. Hettner (1927) transformed this into "chorologic", which refers to explanation rather than description.'

However, in the days when Kant was lecturing in Königsberg, history sought to tell a tale of temporal sequences, while geography described the world and its parts in a systematic way. To suggest, as Werlen does, that Kant would have distinguished between taxonomy and science is about as odd as it is to state that the Swedish botanist Carl von Linneaus (1707–78), who created the taxonomic order of plants in species, families (Latin names), etc., did not think of taxonomy as science. A close reading of Kant reveals that, to Kant, geography was not only descriptive but also predictive. We also have to bear in mind that science has changed a great deal since the late eighteenth century. The change in the appellation of geography from a **chorographic** to a **chorologic** science by Hettner and Hartshorne must also be seen as a consequence of general changes in all the sciences, which took place during the nineteenth and early twentieth centuries. As we shall see later, however, Hettner and Hartshorne were still subsequently attacked for presenting geography as a mere descriptive science.

Today it is widely considered to be impossible, and to some extent philosophically untenable, to draw such sharp divisions between the 'sciences' as Kant, in Hettner's interpretation, did. Understanding of geographical situations is always improved when we consider their development over time. Just as there are geographers who wish to study the cultural landscapes of former times without necessarily using their knowledge to illuminate contemporary conditions, so historians should not neglect the study of differences between places. Here the individual sciences overlap each other.

Hettner considered that the biggest difference between history and geography

> does not lie in that geography sets out to study a given time, namely the present, but that in geography the time aspect recedes into the background. Geography does not study development in time as such – although this particular methodological rule is often broken – geographers cut through reality at a distinct point in time and only consider historical developments in so far as they are necessary to explain the situation at that chosen point in time.
>
> (Hettner, 1927, pp. 131–2)

During the period between the death of Kant in 1804 and Hettner's book in 1927, this 'methodological rule' had not only been broken but also almost systematically opposed, since Kant's philosophical guidelines for the subject had been virtually forgotten until Hettner and Hartshorne resurrected them. In the nineteenth century, geographical research was closely associated with the **study of process**, that is, the development of phenomena over time in geographical space.

THE 'CLASSICAL' PERIOD

While Kant gave geography a theoretical justification, it was Alexander von Humboldt (1769–1859) and Carl Ritter (1779–1859) who laid the scientific foundations for geography as a branch of knowledge. These two men had many views in common and were united in their criticism of the casual and unsystematic treatment of geographical data by their predecessors. The similarity of their opinions was not accidental – Ritter regarded himself as a student of Humboldt, and Humboldt described Ritter as 'his old friend'. They differed markedly, however, in temperament and character.

Humboldt

Humboldt was the last of the great **polymaths**. Penck (1928, p. 31) regarded him as a cosmographer rather than a geographer. He mastered a number of disciplines and put all his energy (and fortune) into travel and research in order to understand the whole complex system of the universe. In his great work *Kosmos*, with its subtitle *Sketch of a Physical Description of the World*, he attempted to assemble all the contemporary knowledge of the material world. *Kosmos* was published (1845–62) in five volumes at the end of his life, while the results from his long and important journeys in Latin America (1799–1804) had been published in 30 volumes in Paris (1805–34).

Humboldt had a gift for exceptionally sharp observation, and in **fieldwork** he was unsurpassed. His books and letters demonstrate that nothing escaped his observation. This, together with his training in geology, is shown in a letter he wrote to a Russian minister after a visit to the Urals: 'The Ural Mountains are a true El Dorado,' he wrote, 'and I am confident, from the analogy they present to the geological conformation of Brazil, that diamonds will be discovered in the gold and platinum washings of the Urals.' A few days later, diamonds were actually discovered (Tatham, 1951, p. 55). Humboldt was primarily interested in the natural sciences. He undertook a large number of altitudinal measurements during his travels and drew height profiles of the continents. These measurements were used in his studies of climatic and vegetation zones which were of basic importance for later research in biogeography and climatology. He also wrote regional accounts of Mexico and Cuba that are still of interest today (Schmieder, 1964).

In contrast to most other natural scientists of his day, Humboldt was not so much interested in discoveries of new species and in collections for natural science museums as he was in understanding the connectivities in the world of nature – a scientific approach he called *physikalische Geographie*. This must not be confused with the more limited concept of '**physical geography**' as used today. The relations between the vegetable and animal kingdoms, and also humankind, with geographical conditions of climate, topography and altitude were central to Humboldt's interests. In his search for *Oeconomia Naturae*, we find some of the roots of modern ecological thinking. His work is much freer of the **environmental determinism** than the much later studies of Ratzel and Semple (see below).

Humboldt transferred Kant's idea of a threefold division of all the sciences to the natural sciences, dividing them into

1) **physiography** or description of nature;
2) *historia telluria*, the geological history of the earth; and
3) *geonosia* or *physikalische Geographie* (Beck, 1982, p. 90).

Humboldt did not wish to confine his research to any one of these subdivisions. As a **cosmographer**, he was not interested in becoming a specialist in either history or geography. Geography and the natural sciences were not then institutionalized and specialized **disciplines**. Humboldt and Ritter would barely have understood the later division of science into disciplines, the specialized fields of work of scientific institutions or the interminable discussions about the nature or limitations of specific disciplines. They were both, in this sense, cosmographers, although their field of inquiry differed.

Ritter

Ritter's recognition as a geographer dates from the publication (in 1817–18) of the first two volumes of his *Erdkunde* – with a volume on Africa and a volume on Asia. The second edition was published in 19 enlarged volumes in 1822–59. On the initiative of leading politicians he was established in 1820 as the first professor of geography in Berlin. Although Ritter travelled widely in Europe, he spent relatively more of his time in his study than did Humboldt. The two men can be considered as complements to each other, in that Ritter was chiefly concerned with studies of human geography, and stood somewhat apart from the rapidly advancing research front of the natural sciences. It would be wrong to describe him as a crude **determinist** as some commentators have done. He believed, as did Vidal de la Blache much later, that 'earth and its inhabitants stand in the closest reciprocal relations, and one cannot be truly presented in all its relationships without the other. Hence history and geography must always remain inseparable. Land affects the inhabitants and the inhabitants the land' (cited in Tatham, 1951, p. 44).

This implied that the individual region or continent had a unity, a *Ganzheit* (which may be translated as a 'whole'), which it was the task of the geographer to study. This entity was something more than the sum of its parts – more than the totality of topographical, climatic, ethnic and other circumstances. Ritter's views were shaped by his deeply religious outlook and by the accepted natural philosophy of his time. His ideas on the 'wholeness' of things were in accordance with the writings of the German 'idealist' philosopher Georg W. F. Hegel (1770–1831), whose attitudes amounted to an attempt to comprehend the entire universe, to know the infinite and to see all things in God (Chisholm, 1975, p. 33). **Holistic** views dominated scientific thought throughout the nineteenth century, and Ritter's ideas were in line with accepted biological, geographical and geological discussion.

The scientific stance of Ritter, like Immanuel Kant before him, was **teleological** (Greek *teleos* = 'purpose'). Teleology seeks to understand events in relation to their underlying purposes. Teleological explanations are therefore often regarded as the opposite of **mechanical** explanations, where the phenomena and observations are understood as outcomes of prime causes such as the 'laws of nature' (see p. 39). Ritter studied the workings of nature in order to understand the purpose behind its order. His view of science sprang from his firm belief in God as the planner of the universe. He regarded the earth as an educational model for humanity, where nature had a God-given **purpose** which was to show the way for humanity's development. He did not regard the shape of continents as accidental but rather as determined by God, so that their form and location enabled them to play the role designed by God for humankind's development.

Ritter's approach to knowledge has been strongly criticized; Hettner (1927, p. 87) said that Ritter's views were in accordance with the spirit of his time, and this reinforced their influence on his generation. This influence, however, was bound to decline when Darwin's *Origin of Species* inaugurated a new philosophy of science.

Ritter combined a basic **teleological** standpoint with a very critical scientific precision: 'My system builds on facts, not on philosophical arguments', he said in a letter (cited in Tatham, 1951, p. 46). The collection of facts was not an end in itself; the systematization and comparison of data, region by region, would lead to a recognition of unity in apparent diversity. God's plans, which give purpose and meaning, could only be discovered by taking into account as objectively as possible all the facts and relationships in the world. In this he followed the advice of Immanuel Kant that the scientist must, on the one hand, continue to explore the universe along mechanistic lines, and should never abandon this approach yet, on the other hand, should bear in mind that organic life can best be understood by minds like ours in terms of the principle of natural teleology (Greene, 1957, p. 68).

Ritter's significance as a scientist lies in his thorough and critical study of sources and in his ability to systematize extensive material. In these respects his work is in the same tradition as that of Humboldt. While earlier regional descriptions had consisted to a large extent of the accumulation of unsystematized data about particular places, Humboldt and Ritter effected a clear structuring of such material and, through deliberate research into both the similarities and the differences between countries and regions, they sought to compare the different parts of the world each with each other.

Ritter was, according to Hassinger (1919, p. 66), 'the founder of the comparative method in regional geography, the victor over every dead aggregate presentation of geographical knowledge. He gave our science unity and coherence' through his use of historical development as its leading thread. Ritter himself, however, regarded Humboldt as the founder of the **comparative method** (Schültz, 1980, p. 44). The basis for Ritter's comparative studies was the physical layout of the land – the area and character of river basins, for example. In this way he could present a land classification or a set of physical regions. The most important part of his work, however, would be to explain the relationships between humanity and nature in the different regions through the history of humankind. Ritter believed that thorough research into the historical development of culture and its relations to physical regions would demonstrate the physical layout of the land as God's educational home for humankind. He therefore made historical development in geographical space the main theme of his educational endeavours. For a long time afterwards geography followed this tradition, including during the period of its **institutionalization** as a specific discipline.

Subsequent writers have commented on the differences between Humboldt's and Ritter's religious and scientific outlooks, and have somewhat exaggerated them. Both men laid great stress on the unity of nature. They both believed that the ultimate aim of research was to clarify this unity and, in this respect, were in accord with the **idealistic philosophies** of their time. Humboldt did not pursue **idealism** in the same way as Ritter, for his concept of the unity of nature was more aesthetic than religious. In this respect he had more in common with Goethe than with Ritter. Unlike Ritter, he saw no reason to explain unity and order in nature as a God-given system to further humanity's development. Humboldt was very much engaged in the gradual development of natural science, and his greatest contributions lay in the field of **systematic** physical geography. Many regard him as the founder of biogeography and climatology. Ritter was, on the other hand, to a considerable extent a **regional** geographer.

The geography Ritter and Humboldt represented was designated as '**classical**' by Hartshorne (1939) because it dominated the foundation – or perhaps we should rather say the prefoundation – period of the subject, since the discipline's institutionalization came much later.

FROM COSMOGRAPHY TO AN INSTITUTIONALIZED DISCIPLINE

No geographer, before or since, has enjoyed a more central position in science and society than Humboldt and Ritter. It might be thought that, in Germany at least, this would have led to a real breakthrough for geography as a scientific discipline. Quite the contrary: with the death of both Humboldt and Ritter in 1859 the 'classical' period came to an end; there were no real successors. The personal professorship Ritter had held ceased with his death, and new university chairs in geography were not established until much later.

Geographical societies

An important reason for this lack of progress was the cosmographic character of Ritter's and Humboldt's achievements. At this time, 'geography' was still not related to a specific discipline; it remained an umbrella concept for a variety of expeditions and other activities within the natural and social sciences, to a large extent supported by **geographical societies** (see Box 2.3) which were flourishing at this time.

Box 2.3　Geographical societies as promoters of expeditions and national interests

The founders of the *geographical societies* included prominent members of the middle classes and enthusiastic scientists who sought to widen support for expeditions and research. These societies also received help from governments during the turn-of-the-century period of colonial expansion. Although forerunners of these societies existed during the sixteenth and seventeenth centuries the first modern geographical society was founded as the Société de Géographie de Paris in 1821. In rapid succession came the Gesellschaft für Erdkunde zu Berlin in 1828, the Royal Geographical Society in London in 1830, then societies in Mexico (1833), Frankfurt (1836), Brazil (1838), the Imperial Russian Geographical Society in St Petersburg (1845), and the American Geographical Society in 1852. By 1885 nearly 100 geographical societies (with an estimated membership of over 50,000) were spread across the world (Freeman, 1961, pp. 52–3).

These societies' most important work was their support for expeditions and their publication of yearbooks and journals, which included maps and other material from expeditions. Many of the societies also supported their respective countries' colonial expansion. Some, such as the societies at Nancy and Montpellier, supported local studies in their home areas. At Nancy, barely 10 km from the Franco-German frontier after the 1870–1 war, the Société Géographique de l'Est pursued studies supporting the return of Alsace-Lorraine to France. During this time geographical societies enjoyed a high profile because of their commitment to popular political causes. Their meetings attracted great public interest, especially when well-known explorers came to give accounts of their discoveries and adventures.

The geographical societies' activities were in marked contrast to the development of academic geography at the time. In the 1880s there were fewer than 20 university teachers in geography in the whole world, and there were far fewer university students in the discipline than members of the geographical societies. The universities were, however, reluctant to give geography status as an academic discipline largely because of the **cosmographic** nature of the investigations promoted by the societies. In the latter part of the nineteenth century, the universities promoted the disciplinary division between the sciences. The information brought back by explorers helped to build up the new disciplines at the universities, particularly in the natural sciences. Many scientists who worked in these disciplines depended on the geographical societies for their data-collecting activities, and feared this support would dwindle if geography became established as a separate academic discipline.

Geography becomes an academic discipline

Many academics considered that geography was not a science; it was only sustained by borrowing from others (Schültz, 1980, p. 65). This argument was based on a scientific philosophy that emphasized natural laws and causality (partial-general reasoning – see Figure 1.2), which was gradually replacing earlier **holistic philosophies**. These scientists found it difficult to accommodate a 'cosmographic' geography as a scientific discipline. The year 1859 however, (in which, incidentally, both Humboldt and Ritter died), saw the publication of Charles Darwin's *Origin of Species*, which provided much of the basis for the new ways of scientific reasoning.

Part of the reason for the lack of new chairs in geography was the fact that the discipline, as expounded by Ritter, was considered by influential academics as merely a subdivision of history or, at best, as a subsidiary subject to it. In his zeal to demonstrate that the earth is a school for humanity, Ritter had become more and more interested in its historical development, and his more influential students concentrated to an even greater extent on historical research. In many schools of geography, especially in France, the subject came to be closely associated with history. In Germany by 1850, geography was a widespread school subject, taught in most cases as an auxiliary subject to history. However, the modern pedagogical principles of the Swiss educationalist Johann Heinrich Pestalozzi (1746–1827) gave geography in Germany, Switzerland and France a somewhat more independent role in compulsory elementary education. Geography was here seen as a means to develop children's power of observation, and the education of geography teachers made the development of academic geography possible. By 1875 there were geography chairs at three Swiss universities (Capel, 1981, p. 59).

It was after the Franco-Prussian war of 1870–1 that influential politicians in Prussia came to realize that geography could also serve an important political purpose. Geographical education could be used to reinforce and popularize the idea of the nation-state (a commendable objective at that time), and could also provide people with a better understanding of the economic and political possibilities of world trade and development. To achieve these goals, geography teachers needed to have a better education. For this reason, in 1874, the Prussian government decided to establish chairs of geography in all the Prussian universities. By 1880, ten professors had been appointed. For more or less the same reason (Boxes 2.4 and 2.5), other European countries also established geography as a university discipline.

One of the reasons why France had lost the war of 1870–1 was considered to be the superiority of the German educational system, which led to calls for school reform and the better teaching of geography. During the 1870s, a handful of chairs of geography were established at French universities, the most significant

Box 2.4 Political arguments for school geography in the nineteenth century: the relations between history and geography

While both history and geography could serve to develop **nationalistic** sentiments, the relative position of these two subjects in the educational system largely came to depend on the degree to which either seemed more useful in building up the idea of a **national identity**. In Germany, with its long history of shifting borders and divisions into small states, geographical patterns associated with the German-speaking lands seemed very significant, and so geography was seen as very important. Norway, on the other hand, had developed its educational system during the union of the crowns with Sweden (1814–1905) and it had no disputed borders. Under these circumstances, Norway's national awakening was fostered by teaching about the glorious history of Viking times and about the precious liberal constitution of 1814. Hence history predominated over geography. Finland, however, which had also experienced a union of crowns (in this case with Tsarist Russia), lacked a clearly discernible glorious past and so geography developed as a relatively more important subject. A pioneering research work of great political importance in the Finnish liberation process, the *Atlas of Finland* (first edited in 1899), stressed the uniqueness of the Finnish lands.

Box 2.5 Arguments for school geography today

As early as 1885, Kropotkin suggested that geographical education is ideally suited to promote a sense of mutual respect between nations and people. Geographical knowledge of other cultures, civilizations and ways of life increases our ability to communicate and is a necessary prerequisite for international co-operation and solidarity. These views have been recently re-emphasized by the Commission on Geographical Education of the **International Geographical Union** in its *Charter on Geographical Education* (1992). This charter particularly addresses the principles of the United Nations, the *World Declaration of Human Rights and the Rights of Children* and the recommendations from UNESCO on education in crossnational understanding, co-operation and peace. The charter points out that geographical education helps to promote the principles of these charters and declarations.

There are four other main reasons why geography should have a role as a 'key discipline' in school education. A 'key' discipline

1) should provide students with some basic concepts to enable them to *understand* the world around them. This is provided by Kant's categories of classification. Space is an elementary and universal key concept. Geography is the science of space and spatial relations;

2) should provide students with qualifications to *cope* with the world around them. Geography teaches the techniques needed for expeditions out in the world. A basic education in geography is needed early in life so as to understand maps, the use of maps, places, peoples and countries. The first task should be to map the path from home to kindergarten or primary school;

3) should explore the possibilities of *synthesis*. Geography is the only discipline that has constantly sought to build bridges between the social and physical sciences, between humanity and nature. Geographical education provides students with the ability to understand the 'wholeness' of a place and how the landscape evolves as a result of natural processes and human actions; and

4) should also provide basic knowledge needed by other *disciplines*. Geographical facts and the understanding of map-making and spatial relations provide essential background knowledge for students of history and the environmental, and social and political sciences.

appointment being that of the young Paul Vidal de la Blache (1845–1918) to the chair at Paris in 1877 (Capel, 1981, p. 60).

Geography at British universities

The political arguments in favour of teaching geography in the universities met with more or less active opposition from within the universities themselves. In Britain, the first personal chair of geography was held by Captain James Machonochie at University College, London from 1833 until 1836. Permanent university teaching was only established from the 1860s onwards when the Royal Geographical Society began to take a growing interest in education. The Society urged the need to establish chairs in geography at Oxford and Cambridge, pointing out that chairs had been or were being established in German, Swiss and French universities. But the proposals were only taken up in 1887, when the Royal Geographical Society offered to cover the main bulk of the costs involved in establishing lectureships at Oxford and Cambridge. This financial support led to the appointment of Halford J. Mackinder to a readership at Oxford in 1887.

Mackinder became an outstanding promoter of geography at both universities and schools. In 1893, together with a group of schoolmasters, he founded the Geographical Association, which did much to encourage and improve geographical education, and although he was elected to Parliament in 1901, he remained an important father-figure in British geography. Mackinder firmly believed that geography should help to train future citizens for the 'world stage' upon which, in Mackinder's view, the British Empire was a force for good. Geographical teaching should therefore be 'from the British standpoint':

> This is, no doubt, to deviate from the cold and impartial ways of science. When we teach the millions, however, we are not training scientific investigators, but the practical and striving citizens of an Empire which has to hold its place through the universal law of survival through efficiency and effort.
>
> (Mackinder, 1911, pp. 79–80)

The first British university **geography department** was established at Oxford in 1900; but geography only achieved a firm footing at Cambridge after 1908, mainly through the continued financial support of the Royal Geographical Society. During a period of 35 years, the society invested about £7,250 in the Cambridge project and £11,000 at Oxford (Stoddart, 1986, pp. 83–125). 'Hence we have the paradox that in the 1880s and 1890s British geography emerged as a scientific

---

Box 2.6   Elisée Réclus (1830–1905) – the life of an idealist

Réclus was first and foremost an uncompromising idealist. He was expelled from the teacher seminary at Montauban in his very first year because he supported the ideals of the 1848 revolution. When only 20 years old he went to Berlin to study theology, but began to attend the popular lectures of Carl Ritter, which awakened his interest in geography. Returning to France in the autumn of 1851, Réclus resumed his political activity. Resisting the *coup d'état* of Napoleon III in the same year, he was obliged to flee to England with his brother (Dunbar, 1981, p. 155). Réclus then travelled extensively in North and South America, more in order to observe than to do research. Sustained research was, in fact, beyond his means, for Réclus had to make a living from the meagre wages he could earn as a tutor and worker *en route*. In 1857, he returned to France and befriended the leading anarchist, Mikhail Bakunin (1814–76). From that time onwards, Réclus belonged to the inner circle of the secret anarchist association Fraternité Internationale. In 1871, he took an active part in the Paris Commune but was captured during the first days of fighting. He was held in prison for almost a year. A sentence of deportation to New Caledonia was commuted to ten years' banishment as a result of the active intercession from geographical societies and such leading personalities as Charles Darwin. Réclus chose to settle in Switzerland in his exile. Although promised a readership in geography in the Université Libre in Brussels in 1892, the university reneged on the appointment for fear of demonstrations after an outbreak of anarchist violence in France in the autumn of 1893. A support committee for Réclus started to collect money and eventually founded the New University of Brussels where Réclus was professor for the last years of his life, refusing to take any salary since his modest needs could be satisfied by the income from his books.

---

discipline without the benefit of aid from strictly scientific men' (*ibid.*, p. 67).

The reluctant and somewhat hostile attitudes in the universities towards the establishment of geography as an academic discipline were largely due to the rapid growth of the systematic sciences and the increasing need to specialize in scientific work. The image of geography as an all-embracing cosmographic subject did not fit in with academic development at this time. Influential academics argued that the fields of geographical inquiry were already covered by existing disciplines. The historian, Edward A. Freeman, could not understand how geography could be recognized as an independent university discipline when 'on the one hand a great deal belongs to history, but on the other the geologists lay claim to much of its material' (Keltie, 1886, p. 466). McKenny Hughes, a professor of geology at Cambridge, argued that his geology department already did all the necessary lecturing and fieldwork in physical geography. At a university, students should specialize in either history, geology, or zoology, drawing on a knowledge of geography already imparted at school (Stoddart, 1986, p. 70). But Mackinder articulated, in a lecture to the Royal Geographical Society in 1887, the need for a geographical science to bridge the gap between the natural sciences and the study of humanity 'which in the opinion of many is upsetting the equilibrium of our culture'. He believed that theories of social evolution, based on Darwinian ideas, would replace the arguments of Ritter (Livingstone, 1992, pp. 190–92; see also Box 2.7, p. 38).

The Keltie Report of 1886 (p. 71) was a powerful and comprehensive documentation of the academic and practical value of geography:

> It is through geography alone that the links can be seen that connect physical, historical and political conditions; and it is thus that geography claims the position of a science distinct from the rest, and of singular practical importance . . . scientific geography may be defined as the study of local correlations.

The USA
In the USA, university teachers of geography included William Morris Davis (1850–1934), who began teaching at Harvard in 1878. The first geography department was established in Chicago in 1903. By the First World War, the discipline was firmly established at a number of North American universities, notably Harvard, Yale, Teachers College (Columbia University), the Wharton School (University of Pennsylvania) and Chicago.

The 'new' geography
The **institutionalization** of geography meant that examinations with national standards had to be organized and the discipline defined as a subject area with a distinct content. This 'new' geography, for which syllabuses and reading lists were drawn up, led to the **professionalization** of people calling themselves geographers. New professional journals were founded, such as the *Geographical Journal* in Britain, *Annales de Geographie* in France and the *Annals of the Association of American Geographers* in the USA.

The new academic geography gradually distanced itself from the geographical societies as university geographers developed a scientific base which the societies lacked. This divergence was, in some cases, widened by the attitudes of prominent members of the societies. The chairman of the Finnish Geographical Society, for instance, maintained the cosmographic view that geography was a collection of different sciences but not a science in itself. This view was of course strongly repudiated by the Finnish professors of geography (Granö, 1986). In the new academic journals (which had often been founded almost in opposition to the geographical societies' journals), geographers set out to justify the new science: they defined its long traditions, its utility and special methods, and delimited geography unequivocally from neighbouring disciplines. A more important function of the academic journals was to publish research that students could use as exemplary models (see Chapter 3).

The new geography drew on some of the ideas of Humboldt and Ritter but was also influenced by developments within the natural sciences which followed the Darwinian breakthrough, and by the political milieu of the period.

THE SOCIAL ANARCHISTS

Réclus
The French geographer, Elisée Réclus (1830–1905), attended some of Ritter's lectures, but was much more influenced by the new political and scientific develop-

ments of his time. He achieved recognition with a work of systematic physical geography called *La Terre* (1866–7), but is best remembered for his 19-volume regional geography *Nouvelle Géographie universelle* (1875–94). The clarity and accuracy of this work made it much more popular than Ritter's *Erdkunde*, which had been its exemplar in many respects but which also contained many obscure passages. Réclus's work became a model for a range of encyclopaedic studies of the geography of the world and of particular countries.

Réclus (see Box 2.6) was probably the most productive geographer of all time. Right at the end of his life he wrote *L'Homme et la Terre* (mostly published posthumously in 1905–8), which may be described as a **social geography** comprising a broad-brush historical presentation of humanity's life on earth and its use of its resources. Réclus also wrote travel handbooks and articles. Although he was the best-known French geographer of his time, Réclus never held a university chair in France and had to earn a living from his writings and, although most of his books were published by Hachette in Paris, he was obliged to live abroad in exile because of his political activities and hence could not directly influence the development of French academic geography.

Réclus's books, including his major work, *Nouvelle Géographie universelle*, showed little evidence of his political beliefs. For this reason they attracted a large, middle-class readership and provided politicians and leading members of the geographical societies with a model for the geography they wished to introduce as a subject in schools and universities. So Réclus had an important indirect influence on the **institutionalization** of geography at a time when most academics saw geography only in its role as a natural science. Like Ritter, Réclus was particularly interested in the human aspects of geography. He had a sharp eye for the inequalities of the human conditions around the world, and made this a central theme of his books. He was also the first European geographer to support the ideas of the American geographer George Perkins Marsh. Marsh had written *Man and Nature, or Physical Geography as Modified by Human Action* in 1864, describing his book as 'a little volume showing that whereas Ritter and Guyot [a student of Ritter who emigrated to the USA] think that the earth made man, man, in fact made the earth' (cited in Beck, 1982, p. 148). Marsh suggested that humanity had, throughout history, destroyed many natural resources and landscapes and he called for a better management of natural resources. In his very first work, *La Terre* (1866–7), Réclus acknowledged his debt to Marsh.

Réclus's political views were more directly expressed in his last work, *L'Homme et al Terre*, which introduced the concept of **social geography**. Since Réclus devoted his political life to social justice, social conditions were inevitably always discussed in his books. He had made a special study of social geography and his inclusion in a travel guide of a description of poverty and relief among the poor of London broke new ground (Beck, 1982, p. 135). Réclus also established a connection between geography and modern town planning and sociology. He influenced and had close contacts with Frédéric le Play, the French sociologist, and Sir Patrick Geddes, the Scottish biologist, social scientist and planner. Geddes, although no anarchist, became a close friend of Réclus in the last decade of Réclus's life. He spread Réclus's ideas in Britain and was most interested in his ideas on social geography, which he found a suitable basis for the development of his work on applied research and planning (see p. 53).

Kropotkin

We have laid emphasis on the life and work of Réclus to demonstrate that the old masters still have something to say to us today. The urge for **social relevance** in geographical work is not a recent phenomenon; it was present in the life and work of the anarchist geographers one hundred years ago. Today Réclus and his fellow anarchist friend and geographer, Count Peter Kropotkin (1842–1921), have aroused renewed interest. The Russian nobleman became famous for his research into the physical geography of northern Europe and Siberia and was a welcome speaker among the geographical societies of western Europe. Kropotkin also gave Réclus ample help with the sections of *Nouvelle Géographie universelle* that dealt with Siberia and eastern Europe.

Kropotkin voiced his political opinions much more directly than Réclus, trying to revolutionize the discipline of geography in a number of ways. He presented theories of geographical education, the relations between humanity and nature, and of decentralization, which were clearly aimed at discouraging the use of geographical research for exploitative and imperialistic purposes (Breitbart, 1981, p. 150).

Kropotkin's political ideas had developed while he was travelling in Siberia during the 1860s. He was impressed by the Russian peasants' spirit of equality and self-sufficiency but was disheartened by the negative effects of political centralization and social inequality. In 1871, Kropotkin became an active advocate of **social anarchism** and, as a consequence, had to go into exile. From that time onwards he devoted his life to two revolutions: one in social and economic relations, the other in the discipline of geography itself. Kropotkin's social anarchism sought to demonstrate that people had a capacity to develop a better society on a co-operative basis, provided that the structures of domination and subordination were taken away. He believed that centralized institutions inhibit the development of a co-operative personality, promote inequality and limit economic progress (*ibid.*, p. 138). He advocated small-scale economic activity within and between regions which, he thought, could produce economic development superior to that produced by large-scale industry. Such arguments are advanced today by the 'green' movements and by economists such as Schumacher in *Small is Beautiful* (1974). Kropotkin suggested that large city regions should be disaggregated into small, more or less self-sufficient townships within which living, working and recreational space might be integrated. Such ideas were later put into practice by the **'garden city' movement** led by Ebenezer Howard and Lewis Mumford. Another important connection was to Patrick Geddes (see pp. 53–54).

Most significantly for his time, Kropotkin opposed strongly the dominant interpretation of Darwin's *Origin of Species* (1859), which saw nature as an immense battlefield upon which there is an incessant struggle for life and the extermination of the weak by the strong. He attacked the **social Darwinist** views of the influential philosopher Herbert Spencer (1820–1903). Spencer envisaged human societies as closely resembling animal organisms which must engage in a constant struggle to survive in particular environments. Spencer believed that the 'fittest individuals' would survive best in a free-enterprise system, and thus lead civilization forward. In his 'general law of evolution' he claimed that all evolution is characterized by concentration, differentiation and determination. Kropotkin, on the other hand, wrote (1924, p. vii): 'I failed to find, although I was eagerly looking for it, that bitter struggle for the means of existence between animals of the same species'. The struggle for existence may be hard, but it is not carried out by individuals – rather, by groups of individuals co-operating with each other. In the development of civilization, as in the quest for survival, Kropotkin believed that mutual aid within small, self-contained communities was the best workable solution.

Both Réclus and Kropotkin stressed and developed the idea that geography as a discipline should encompass both humanity and nature. Kropotkin was a close friend of Scott Keltie (the secretary of the Royal Geographical Society) and strongly supported the arguments that were used to establish geography as a university discipline in Britain (the 'Keltie Report' of 1886). But for Keltie it was essential that science should be politically and socially neutral and value free – that is, based on the scientific ideals of **positivism** (see Chapter 4). In writing the obituary of Kropotkin, Keltie (1921, p. 319) expressed the view that 'this is not the place to deal in detail with Kropotkin's political actions, except to regret that his absorption in these seriously diminished the services which otherwise he might have rendered to geography'. Réclus and Kropotkin may have had little direct influence on the geographical establishments of their day but they retained alight a flame of critical social inquiry that has been resurrected today.

DARWINISM

Kropotkin's ideas demonstrate that, during the late nineteenth century, it was considered important that we learn how to shape human societies from nature. **Darwinism**'s influence was strong, not only on social thinking but also on the development of science as such. There are good reasons to purport that, alongside the Bible, no book has had a more profound influence on our ways of thinking than Charles Darwin's *Origin of Species* (1859). Stoddart (1966) suggests that the following four main themes from Darwin's work can be traced in later geographical research:

1) Change through time or evolution – a general concept of gradual or even transition from lower or simpler to higher or more complicated forms. Darwin used the terms 'evolution' and 'development' in essentially the same sense.
2) Association and organization – humanity as part of a living ecological organism.
3) Struggle and natural selection.
4) The randomness or chance character of variations in nature.

The concept of change over time is, for instance, found in Davis's cyclic system of the development of a landscape through the stages of youth, maturity and old age, which Davis himself described as evolution. The French school of regional geography (pp. 45–48) also used the term in their study of the changing cultural landscape.

Box 2.7   Evolution and society

The philosophical reflections of the Greek and Roman geographers on the relationships between humanity and the environment survived into modern times. The belief that the earth had order and purpose designed by a deity had been fundamental to Ritter's approach, but did not survive into the latter part of the nineteenth century. The view that the environment had an influence on people and on the development of human society became central to **environmental determinism**. A more positive alternative scenario was **possibilism**, which studied the extent to which people were able to modify their environment. Possibilists also recognized that humankind could have negative influences on the environment (as, for instance, suggested by Marsh and Réclus.

   The idea of **evolution**, the gradual or even transformation from lower and simpler to higher and more complicated forms of life, dominated scientific debate in the latter part of the nineteenth century, in both the natural and social sciences. There were fierce debates on whether evolution was strictly governed by inheritance and selection of the fittest through random biological variations, or whether qualities acquired or learnt by an organism during its lifetime could be passed on to its offspring. The latter concept had been promoted by Jean-Baptiste de Lamarck (1744–1829), a gifted French natural scientist who lived before Darwin's time.

   Livingstone (1992, p. 187) maintains that geography was influenced more by the social ramifications of the neo-**Lamarckian** version of evolutionary theory than by Darwin's own ideas. Neo-Lamarckian arguments support the belief that social progress can be greatly accelerated by learning and through the acquisition of habits in response to adaptations to the environment. Biologists eventually agreed that Lamarck had been wrong; an individual's acquired abilities are not passed on to the next generation by inheritance. Today, of course, social theory no longer emphasizes the laws of natural science: social environment may be more important than biological inheritance in shaping life for future generations.

   Social Darwinism became a catch-all term used to characterize a variety of social theories that emerged in the latter part of the nineteenth century (*ibid.*, p. 184). These theories have particularly been associated with the influential philosopher, Herbert Spencer (1820–1903). Spencer's ideas buttressed ***laissez-faire*** individualism in economic and political life. He believed that the development of civilization was a manifestation of the survival of the fittest, an argument much favoured by big business and imperialists. For instance, John D. Rockefeller, the founder of Standard Oil, saw the growth of big business as the outcome of natural selection.

   On the other hand, political liberals used Darwinian language to make their assault on aristocratic privilege and landed property, which could protect some idle and unproductive members of society from competition with more entrepreneurial individuals. The writer George Bernard Shaw, for instance, suggested that the quality of civilization might be improved by **eugenics**, the 'science' of improving the qualities of the human race by the careful selection of parents. Such ideas developed into Nazi racial ideology, as later voiced by Hitler.

The ideas of association and organization – which Darwin had inherited from earlier philosophers and scientists – have been rather tenacious in geographical research. The influential German geographer, Friedrich Ratzel (1844–1904), discusses in his *Political Geography* (1897) 'the state as an organism attached to the land'. Although this **organism analogy** was derived from the natural sciences, its roots may also be traced back to the earlier idealist philosophy and concept of *Ganzheit* or 'whole' as used by Ritter (see above). The **region** (that particular field of study for geographers) has been regarded as a unique functional complex which, despite a steady stream of material and energy, is in apparent equilibrium and constitutes a 'whole' that is more than the sum of its parts. This idea of the region is to be found in the French school of regional geography, and is also apparent in relatively more recent university textbooks (Broek and Webb, 1973, pp. 14–15). The concept of the region as an organic unity pervaded British geography in the first half of this century. Andrew John Herbertson (1865–1915) used the term 'macro-organism' for the 'complex entity' of physical and organic elements of the earth's surface (Stoddart, 1966, p. 691).

The concept of struggle and selection, which had parallels in the contemporary political ideas of both economic liberals and Marxists, is reflected in subsequent geographical research, although no agreement was reached as to whether variation and development occurred by chance or were predetermined. This reflects a certain ambiguity in Darwin's thought. In later revisions of his work Darwin abandoned the issue of random variation and chance, partly because he had failed to discover the laws which he had earlier believed to govern chance variations, and partly as a response to churchmen who sought to reconcile evolutionary process with fundamental direction (Box 2.7).

The scientific method that dominated the latter part of the nineteenth century was in marked contrast to the **teleological** thinking of the 'classical' period, which had supported and sustained the pursuit of a cosmographic science. It was no longer science's objective to be a witness to the existence of God or to find in His grand plan, **final causes** or purposes in what was observed. Scientists sought to determine the laws of nature as **prime causes** which might explain observed reality. This change in viewpoint gradually led to the replacement of **inductive** reasoning by **hypothetic-deductive reasoning**. Hypothetic-deductive reasoning sets up hypotheses to see how far they can explain observed reality. If and when the testing of these hypotheses yields positive results, scientists are on the way to formulating **scientific laws** (see Chapter 3). Scientists were now in general agreement that religion could not provide explanations for natural phenomena: some theologians were even prepared to accept that the Bible was not an authoritative source on scientific matters. Discussion as to whether or not there was a Thought or a God behind the laws of nature, or whether natural laws were the means of development towards a designed end, came to be regarded as unscientific. Questions of belief and questions of knowledge now occupied two different realms, with science concerning itself with causes rather than purposes. This increasing belief in the inherent value of science led to a considerable growth in research into the natural sciences.

GEOMORPHOLOGY AND PHYSIOGRAPHY PROVIDE
ACADEMIC RESPECT

During the latter half of the nineteenth century, it was primarily a group of
natural scientists with interests in physical geography who won the subject academic respect. A start had been made in Britain by the first prominent female
geographer, Mary Somerville (1780–1872), whose *Physical Geography* had been
published in 1848. This book gave detailed descriptions of the topography of each
continent (without illustrations) and of the distribution of plants, animals and
human beings. It was not a simple physical geography since it included the works
of human beings, being more an exercise in the **cosmographic** tradition.

The German school of geomorphology
As a result of Darwin's work, **geology** and **biology** became the most ambitious
branches of the natural sciences. Geology was important because it could, with
the aid of a **palaeontology** (which interprets fossils), clarify the evolution of plant
and animal species. These sciences had been elements within cosmographic geography and their independent development (which led to the establishment of
university chairs) was a threat to the universalistic claims of geography. Some
opportunities were created for geography, and these were first appreciated by the
German, Oscar Peschel (1826–75). In an otherwise scientifically weak book, called
*New Problems of Comparative Geography as a Search for a Morphology of the Earth's
Surface* (1870), Peschel chose the right moment to propose that geographers
should study the **morphology** of the earth's surface. Like Ritter, he was interested
in the significance of landforms for the development of human beings, but he did
not share Ritter's religious outlook, being more concerned with causes and effects
as illustrated by the methods of the natural sciences.

   Geologists had set themselves such huge tasks – including the development of
geological timescales, the systematic mapping of rock types and the analysis of
fossils – that they found it difficult to cope simultaneously with the morphology
of the earth's surface. Here was a field where geography could carve out a place
for itself as a scientific discipline. The study of landforms became the leading field
of research for most of the professors appointed to geography chairs in the latter
half of the nineteenth century. The biological elements in nature subsequently
came to play a relatively small role in the teaching of geography in most countries,
making **physical geography** a more appropriate description of what was taught
than the term **geography of nature**.

   One important reason for the 'geologification' of geography during the 1870s
and 1880s was that the new professors needed academic qualifications, preferably
doctorates. Friederich Ratzel (1844–1904) said: 'they told me when I returned from
my expeditions that they needed geographers'. So he organized as quickly as
possible all the material he had collected on the Chinese migration to the USA and
wrote it up as a doctoral thesis, which gained him the chair of geography at
Munich in 1875 (Jean Brunhes, 1912, after a discussion with Ratzel in 1904). There
were, however, many academics who had a scientific training in geology. Ferdinand von Richtofen (1833–1905) was a well-known geologist and explorer when
he was offered the chair of geography in Bonn. Although he did not want to

exclude humanity from his teaching of geography, he made the study of land-forms the main research field for himself and his students.

Even more influential in this development was Albrecht Penck (1858–1945), a dominating figure in German geography. Internationally, Penck's influence could only be compared with that of the American, William Morris Davis (1850–1934), with whom he was in close contact – despite their very different interpretations of the **cycle of erosion** (Martin, 1985, p. 224). Penck, like von Richtofen, was a geologist by training, but he was primarily interested in the **Quaternary period** and worked on the identification of **glacial periods** and on **glacial morphology** in general. Early in life, Penck developed his glacial theory, which he used to explain many of the landforms of central and northern Europe. His main work (1901–9) was on the glacial periods in the Alps and their effects on Quaternary sedimentation. His identification of the four **glacial periods** (Günz, Mindel, Riss and Würm) was a major step in the development of glacial geomorphology. Having been advised by a geologist friend to take his doctorate in geography, Penck was, at the age of 27, appointed to a chair of physical geography in Vienna. He founded an influential school of geography there but in 1906 moved to Berlin, where he held an influential position until his death.

Penck's work had a major impact in the Scandinavian countries and, for a long time, Scandinavian academic geographers were primarily concerned with the study of the effects of glacial periods on the landscape. The leading professors, generally the first to be appointed, were geomorphologists. Another important contributor to geomorphological research was the Serbian geographer, Jovan Cvijic (1865–1927). His work on the **karst phenomena** in the limestone regions of his homeland introduced Serbo-Croat words like *dolina* and *polje* into the scientific vocabulary (Freeman, 1980, p. 19).

Huxley and Davis

In Britain, no work was more influential in changing the teaching of geography at all levels than T.H. Huxley's *Physiography*, first published in 1877. At this time the term '**geomorphology**' had not yet been invented and only '**morphology**' was in general use. '**Physiography**' has a much wider meaning; it may be defined as a 'description of nature', encompassing the systematic sciences of botany, geology and zoology. One reason for the success of Huxley's book was its demonstration of better and more interesting ways of learning, linked directly with the pupil's own experience. The book begins with the Thames at London Bridge and, working from the local and familiar to the unfamiliar, it ends with the earth as a planet. The characteristic of the book is its emphasis on experimentation and local studies. The idea that geography can only be learnt through local studies, field courses and excursions is, to a large degree, derived from the educational principles of the Swiss educationalist, Heinrich Pestalozzi (1746–1827), and it was these principles that were developed into practice by Huxley.

'**Physical geography**' (renamed '**physiography**' after 1877) became a very popular school subject during the last third of the nineteenth century, accounting for some 10% of the examination papers sat in English and Welsh schools during that time (Stoddart, 1975, p. 26). Physiography was now regarded as an integral, if not the most important, part of geography.

As more specialized earth sciences became established, physiography as a comprehensive subject gradually vanished from the syllabuses. Within geography, its place was filled by the new science of '**geomorphology**', which was introduced to Britain in 1895. The American, William Morris Davis (1850–1934) became the leading personality in the development of geomorphology in the English-speaking world, although he preferred the term '**physiography**' himself. His organizing principle was not simple causality but rather the idea of regular change of form through time, as systemized in his scheme of the **cycle of erosion**.

Based on his investigations of valley formations in Pennsylvania, Davis presented his first systematic exposition of the erosion cycle in 1889. This was an ideal typification of landform processes. Initially, a rapid tectonic uplift created a young landscape, which was immediately attacked by the destructive agents of erosion that would, in time, reduce the landscape to its ultimate old-age form, a low flat plain termed a **peneplain**. Davis argued that most places had gone through several complete cycles of erosion, and that present-day high mountain plateaux were merely remnants of former base levels.

The success of physical geography, variously termed '**physiography**', '**morphology**' or '**geomorphology**', led the German Georg Gerland to suggest in 1887 that the study of cultural phenomena should be separate from geography. He supported this view on the logical grounds that geography is the science of the earth, and that a science should be developed on the basis of natural scientific methods. Scientific methods, as defined, could not be used to study cultural phenomena.

Developments in physical geography were the main innovations in the discipline during the latter part of the nineteenth century. Gerland's proposals indicate how far the pendulum had swung in this direction, and his arguments may have led many to conclude that geography had lost its basic, theoretical balance.

ENVIRONMENTAL DETERMINISM AND POSSIBILISM

After Darwin, geographical research was primarily concerned with discovering the laws of nature. Nature was studied with open eyes: geographers sought as objectively as possible to identify the natural processes that governed the formation of valleys, uplands and coastlines. A more restricted view was taken of human activity; only the relationships between nature and humanity were considered to be of prime interest. Humanity's achievements were explained as consequences of survival of the fittest under the pressure of natural conditions (see Box 2.7, p. 38).

Ratzel
Friedrich Ratzel, widely recognized as the founder of human geography, was greatly influenced by such ideas. The first volume of his chief work, published in 1882, was entitled *Anthropogeography, or Outline of the Influences of the Geographical Environment upon History*. In this volume (which sought to transfer the new methods of the natural science to studies in human geography), Ratzel stressed the extent to which humanity lives under nature's laws. He regarded cultural forms as having been adapted and determined by natural conditions. Although

he may be criticized today for his determinism, we should acknowledge that Ratzel broke new ground in demonstrating that cultural, as well as natural, phenomena can be subjected to systematic study. Before his time human geography had largely confined itself to regional studies.

Environmental determinism is also evident in Ratzel's *Political Geography* (1897). In this book, the establishment of states is seen as an evolutionary necessity, and Ratzel explains the growth of political unities in terms of organic growth (Uhlig, 1967). These ideas quite clearly influenced his Swedish pupil Rudolf Kjellén, who was to become a propounder of **geopolitics**. Kjellén simplified and popularized Ratzel's ideas, maintaining that states follow the principle of the 'survival of the fittest' and have an independent existence over and above that of their citizens. These ideas were readily accepted and followed up by Karl Haushofer (1869–1946) who, as a geographer, tried to build up geographical arguments for a German right of domination over neighbouring peoples. Geopolitics became an important part of Nazi ideology. While it is unfair to hold Ratzel responsible for this much later development, we must be aware of the connections between environmental determinism, nationalism and racism (Peet, 1985). Somewhat the same, maintains Livingstone (1992, pp. 194–6), can be said of the founding father of British geography, Halford Mackinder. Based on his ideas of social evolution, Mackinder developed a political geography, that greatly influenced British foreign policies and military strategies. Thinking geographically was to him part of a strategy to secure the 'maintenance and progress of our Empire' (Mackinder, 1911).

In the second volume of *Anthropogeographie* (1891) Ratzel discussed the concentration and distribution of settlement forms, migrations and the diffusion of cultural characteristics. In this work he was influenced by his 'fatherly friend', the German evolutionary theorist Moritz Wagner, who argued that Darwin had failed to appreciate the significance of migration and geographical isolation for the process of evolution. Ratzel realized that he could not solely explain phenomena in terms of natural conditions, but that the historical development and cultural background of local and migrant populations were also significant. At one point he even declared (as cited in Broek, 1965, p. 18): 'I could perhaps understand New England without knowing the land, but never without knowing the Puritan immigrants.'

The promotion of determinism

In scientific circles, however, the first volume of *Anthropogeographie* had a much greater impact than the second volume. Ratzel's American disciple Ellen Churchill Semple (1863–1932), in particular, laid great significance on Ratzel's deterministic opinions in her teaching at American universities (Box 2.8, p. 44). Semple and others of the same way of thinking, including the geomorphologist William Morris Davis, dominated American geography until the 1930s and exerted an influence over American school geography for much longer. In summing up the principles of 'geographical evolution' to the readers of a school magazine in 1900, Davis maintained that two great principles had been discovered, both of vast importance for geography. One was the evolution of landforms, contributed by geology; the other, the evolution of living forms, contributed by biology (Livingstone, 1992, p. 205).

Box 2.8   Ellen Churchill Semple on environmental influences

A major work by the American geographer Ellen Churchill Semple is entitled *Influences of Geographical Environment* (1911). It begins (p. 1) as follows:

> Man is a product of the earth's surface. This means not merely that he is a child of the earth, dust of her dust, but that the earth has mothered him, fed him, set him tasks, directed his thoughts, confronted him with difficulties that have strengthened his body and sharpened his wits, given him problems of navigation or irrigation, and at the same time whispered hints for their solution.

Semple believed that human temperament, culture and economic life could all be the result of environmental influences. She argued, for instance, that 'the northern peoples of Europe are energetic, provident, serious, thoughtful rather than emotional, cautious rather than impulsive. The southerners of the subtropical Mediterranean basin are easy-going, improvident except under pressing necessity, gay, emotional, imaginative, all qualities which among negroes of the equatorial belt degenerate into grave racial faults.' (*ibid.*, p. 620).

Davis, like Ratzel and Mackinder, was determined to keep the study of human and physical geography together under one explanatory umbrella. Given their educational background in the natural sciences and the scientific importance of evolutionary theory at that time, it seems inevitable that they would try to transfer evolutionary concepts to the study of the human world. Livingstone (*ibid.*, p. 210) has argued that the vocabulary best suited for their purposes was provided by the neo-**Lamarckian** version of evolutionary theory (see Box 2.7, p. 38).

Two influential geographers who continued the study of environmental effects until the middle of the twentieth century were Ellsworth Huntington (who related the rise of civilization in the mid-latitudes and the lack of development in the tropics to climatic conditions) and Griffith Taylor, 'whose determinist views so angered politicians interested in the settlement of outback Australia that he was virtually hounded out of his homeland' (Johnston, 1997, p. 43). Due to climatic conditions Taylor insisted that possibilities for new settlements were limited. Taylor moved to the USA in 1928 and later to Canada; he insisted that he was not an old-fashioned determinist, but had based his views on scientific knowledge of the environment. In 1951 Taylor (p. 7) could state with satisfaction: 'Thirty years ago I predicted the future settlement-pattern in Australia. At Canberra (in 1948) it was very gratifying to be assured by the various members of the scientific research groups there, that my deductions (based purely on the environment) were completely justified.' At the age of 70 Taylor returned to Sydney, was welcomed as a national hero and was even commemorated with his picture on a stamp. This example shows that a scientifically based **environmental determinism** might have some merit.

Until the middle of the twentieth century, strong elements of crude environmental determinism played a major role in the selection of topics that were included in school textbooks in most countries. A 'commercial' type of geography was popular that laid considerable emphasis on raw material supplies and little on the distribution of entrepreneurial skills and institutions. The influence of physical geography on transport routes was greatly overstressed and oversimplified.

The advent of possibilism

Geographical research in Germany was soon to react against crude environmental determinism. Alfred Hettner (1859–1941) was the most significant contributor to this. Hettner was the first German to enter university with the declared aim of becoming a geographer. What intrigued him about geography was the idea of humanity's dependence upon nature (Beck, 1982, p. 215). Gradually, however, he modified his environmentalism and, in 1907, declared that as far as we restrict discussion to the influence of nature upon human beings, we are only dealing with possibilities, not certainties. In his monumental presentation of the history, content and methodology of geography, *Die Geographie, ihre Geschichte ihr Wesen und ihre Methoden* (1927), Hettner asserted that geographical synthesis is distorted when nature is regarded as dominant and humanity as subsidiary. Through this book and through his editorship of *Geographische Zeitschrift*, Hettner became the most influential philosopher of geography. He maintained that the primary cultural task of geography is to build a bridge over the gap that had opened during the latter part of the nineteenth century between the natural sciences and the humanities.

Hettner reformulated Kant's concept of **chorography** (the description of phenomena that are found at the same place) to **chorology** – a scientific explanation of the total causal relationships of an assemblage of phenomena that are mutually co-ordinated, but not subordinated, in places. **Regional geography** (*Länderkunde*) was therefore the crucial part of geography. As Hettner moved away from environmentalism, he became more and more opposed to the development of **geopolitics** and particularly to the notion that there is some correlation between race and culture (Schültz, 1980).

The view that 'there are no necessities, only possibilities' was strongly urged by the French historian Lucien Febvre (1922), who termed this approach **possibilism** and contrasted it with environmental determinism. Febvre invented this term, but the development of the possibilist way of thinking had started earlier and is especially associated with the French geographers Paul Vidal de la Blache (1845–1918) and Jean Brunhes (1869–1930). Later, Isaiah Bowman (1878–1950) and Carl Sauer (1889–1975) became active advocates of possibilism in the USA. The possibilists did not deny that there were natural limits to the activities of humanity but emphasized the significance of humanity's choices of activity, rather than the natural limitations to it.

THE FRENCH SCHOOL OF REGIONAL GEOGRAPHY

Paul Vidal de la Blache is regarded as the founder of modern French geography. Vidal started his academic career as a student of classics and history. It is not clear why he transferred his interests to geography, but the fact remains that, when he was appointed to a chair in history and geography at Nancy in 1873 (when only 28), he sought permission to teach geography alone. Andrews (1986) links this change of interest to the transformation of the French secondary school curriculum, which was inaugurated at this time. As a consequence, better geography teachers were needed in French schools and it seems that Vidal grasped this opportunity at the right moment. He became *professeur* and holder of the chair in geography (alone) at Nancy in 1875, when he had attained the age of 30, which was the minimum required for the title. Nancy set the example for the Ministry of Education, who became convinced of the desirability of naming chairs in geography alone rather than following the long-standing tradition of combined chairs in history and geography. Some years later, Vidal moved to Paris where he was to become the father-figure of French geography.

Vidal pointed out the weakness of deterministic arguments, the futility of setting humanity's natural surroundings in opposition to its social milieu and of regarding one as dominating the other. He considered it even less useful to tackle these relationships along systematic lines in the hope of discovering general laws that might govern the relationships between human beings and nature.

According to Vidal, it is unreasonable to draw boundaries between natural and cultural phenomena; they should be regarded as united and inseparable. In an area of human settlement, nature changes significantly because of the presence of human beings, and these changes are greatest where the community's level of material culture is the highest. The animal and plant life of France during the nineteenth century, for example, was quite different from what it would have been had the country not been inhabited for centuries by human beings. Hence it is impossible to study the **natural landscape** as something separate from the **cultural landscape**. Each community adjusts to prevailing natural conditions in its own way, and the result of the adjustment may reflect centuries of development. Each single small community, therefore, has characteristics that will not be found in other places, even in places where the natural conditions are practically the same. In the course of time, humanity and nature adapt to each other like a snail to its shell. In fact, the relationship between humanity and nature becomes so intimate that it is not possible to distinguish the influence of humanity on nature from that of nature on humanity. The two influences fuse. Through time local lifestyles (*genres de vie*) develop an integrated web of physical and social threads.

The area over which such an intimate relationship between human beings and nature has developed through the centuries constitutes a **region**. The study of such regions – each one of which is unique – should be the geographer's task. Vidal, therefore, argued for regional geography and against systematic geography as the discipline's core.

The *méthode vidalienne*

Vidal's method, which was **inductive** and **historical**, was best suited to regions that were 'local' in the sense of being somewhat isolated from the world around them and dominated by an agricultural way of life. These circumstances favoured the development of local traditions in architecture, agricultural practices and general way of life; these communities lived in such a close association with nature that they might be self-sufficient in the majority of goods. Vidal advised geographers to carry out research in folk museums and collections and to investigate the agricultural equipment used in the past in order to study the region's individuality of development.

Vidal (1903, p. 386) used the following illustration to underline the long associations between the major factors governing a community's development: while the surface of a shallow lake is being swept by a gust of wind, the water is disturbed and confused but after a few minutes the contours of the bottom of the lake can clearly be seen again. In the same way, war, pestilence and civil strife can interrupt the development of a region and bring chaos for a while, but when the crisis is over the region's fundamental developments reassert themselves. Changes can occur in such a community, and Vidal pointed out many developments which had taken place in the French regions during the centuries preceding the French Revolution, but these developments had taken place within a stable framework of interaction between human beings and nature.

Vidal's characterization fitted fairly well those virtually unchanged communities in Europe with which he was most familiar. These had been agricultural societies throughout the Middle Ages and well into recent times, and had essentially local interests and interaction patterns resulting from a long interplay between humanity and nature. In certain circumstances, and for certain members of the small upper classes, the field of contact was national rather than local but, in general, Vidal's concepts were appropriate. Ironically, however, he was describing a rapidly disappearing phenomenon.

His method is still well suited to the study of the historical geography of Europe up to the Industrial Revolution, and also for the study of those parts of the world where society depends on subsistence economies. As Wrigley (1965, p. 9) pointed out, however, Vidal's method is not so well suited to the study of regions that experienced the Industrial Revolution of the nineteenth century. Towards the end of his career, Vidal was well aware of this situation. This is most evident in *La France de l'Est* (1917), in many ways his most original work, which studied the development of the landscapes and agricultural societies in Alsace and Lorraine over a period of two thousand years. A considerable portion of the book, which is arranged chronologically, is devoted to changes that preceded the French Revolution of 1789. The revolution produced a great ripple in the picture, but afterwards the main lines of development reasserted themselves. After a particular point in the nineteenth century, however (Vidal identified the year 1846 himself), the finely balanced interplay between humanity and nature was profoundly disturbed. The surface of the lake was, as it were, whipped up by something more significant and persistent than the earlier gusts. The visible contours of the bottom of the lake were reshaped. The building of canals and railways, together with the reformation of the Alsatian woollen industry through the introduction of cotton mills, initiated the decline of the traditional local, self-sufficient economy. Industry was developed on the basis of new, cheap raw materials and rapid means of transport, and hence it could mass-produce goods for a wider market. And it was these developments that reduced the validity of the regional method in a growing number of areas.

Vidal regretted these developments, which he could not avoid observing. He considered that much of the best in French life was vanishing with the self-sufficient economy, but his stature as a scholar enabled him to suggest possible new approaches for further research. In the future, he suggested (*ibid.*, p. 163), we should study the economic interplay between a region and the city centre that dominates it, rather than the interplay of natural and cultural elements. Despite the breakdown of the self-sufficient regional economy, Vidal's work has been and still is a great inspiration to a vital tradition in geography – that of the **regional monograph**.

Regional monographs and the Annales school

In France there was a strong connection between regional geography and the **'Annales' school of history**. _Annales_ is the name of a journal launched by Lucien Febvre and Marc Bloch in the early 1930s. Its aim was to reform the discipline of history by focusing on the history of ordinary people through such means as archaeological and other physical evidence and private archives, rather than through traditional way of recounting history based on a study of written sources that stressed political events, military confrontations and the lives of kings, emperors and ministers. The 'new' history of the Annales school sought to discover the fate of the lower classes, the ways in which they were organized, and the techniques they used in farming or producing tools and other artifacts (Claval, 1996, p. 319).

Regional monographs (following the models of Vidal de la Blache) were concerned with the development of material culture. They were developed further by the Belgian geographer Pierre Gourou, among others. Gourou's first research (1936) was on the peasant culture in the Tonkin Delta of Vietnam. He was fascinated by the farming techniques which allowed an almost perfect control of water in a very difficult environment. He found that it was possible to understand peasant societies like the Tonkinese only through study of their political and social organization, thus deepening the concept of _genre de vie_. As we shall see, this leads on to modern geographical methods in the study of place, landscapes, regions and space (Chapter 5).

The French **regional monographs** acted as models for geographers around the world. At the Institute of Geography at Belgrade University, for instance, under the inspiring leadership of Jovan Cvijic, 25 monographs on the regions of Serbia were produced in the years before the First World War. These investigations made Serbia one of the best-studied countries of the world, and were also instrumental in the creation of the state of Yugoslavia at the Paris Peace Conference in 1919. In Britain, regional monographs were widely admired. They were studied and produced in the Oxford School of Geography and later introduced into the newer universities by Percy M. Roxby and others (Buttimer, 1983, p. 91).

LANDSCAPES AND REGIONS

Vidal de la Blache is only one, if the most significant, of the founders of regional geography. There are other directions and methods in regional geography, although their discussion and classification have been rather neglected by English-speaking geographers. In Germany and France (where regional geography has been regarded as the core of the subject until quite recently), contributions have been much more extensive in this field. Learning in Germany has a long tradition of classification and systematization, and we can turn to Fochler-Hauke's _Geographie_ (1959, pp. 251–61) for a review of the different approaches to regional geography.

Before we do so, we must first clarify the concepts 'region' and 'landscape' and their German counterparts. The two German words _Land_ and _Landschaft_ may both be translated as 'region', but _Land_ is a definite unit – a county or a country normally deemed by its administrative borders. _Länderkunde_ is the art of describing such definite units, as in regional monographs such as Demangeon's _La Picardie_ (1905) – regional geography in its traditional sense. The word _Landschaft_ can be used, as in

English, to refer to the *landscape* scenery of an area since there are many similarities in terms of the origins and the development of the German and English words. The German word, however, has a more explicit territorial meaning than the modern English word, in part because it has, historically, been used to refer to a small, often relatively independent, territory. German geographers have given this a more specifically scientific meaning: a small region that is defined, in part, by the character of its physical features. The difference between the English emphasis on landscape as scenery and the German emphasis on territoriality has led to differences of opinion regarding the appropriateness of using the German term in English (for a more thorough discussion, see Olwig, 1996). The German science *Landschaftskunde*, which concerns both the study of such small unique areas and the delimitation and classification of different types of regions, straddles regional and systematic geography. Hannerberg (1968, pp. 132–3) considers that *Landschaftskunde* could be taken over by the systematic branches of geography, while *Länderkunde* remains as regional geography in its own right. With this in mind, we can now turn to the five approaches to *Landschaftskunde* as distinguished by Fochler-Hauke: landscape morphology, landscape ecology, landscape chronology, regionalization and landscape classification (or systematization).

Landscape chronology

With its concern with development over time as a device to present regional synthesis, **landscape chronology** might seem to include French regional geography. However, this approach must be regarded as a method of regional geography or *Länderkunde*, and not as an approach to *Landschaftskunde* (see Figure 1.5, p. 10).

Landscape chronology concerns particularly the specific scientific work of reconstructing former landscape types – for example on the basis of relict elements in the contemporary landscape. The use of aerial photographs to detect the ridge and furrow patterns of ancient fields (as first analysed by Mead (1954) and in the study of the lost villages of medieval England (Beresford, 1951)) is an example of the research that might be used to reconstruct a chronology of shifting landscape types through history. As a research technique this might seem rather thematic and restricted, but it is possible to make the study of shifting landscape types the leading thread in a fully fledged regional presentation.

Landscape chronology might also include the type of geographical study described by Derwent Whittlesey (1929) as **sequent occupance**. Sequent occupance concerns the ways in which each culture uses a region in its own particular way. This can be demonstrated in America where most regions experienced a sudden change from Indian to European cultures, and also in many parts of Europe which progressed from agrarian to industrial cultures. Sequent occupance stresses the stages in the development of a region and not, as in the France of Vidal de la Blache, local differentiation as a result of the long-continued and largely undisturbed interplay over the centuries of humanity and nature.

Landscape ecology

Although particularly well developed in Germany, the concept of **landscape ecology** is less understood in the Anglo-American world. Carl Troll (1899–1975), who introduced the concept, defined landscape ecology as the complex of causal and reciprocal connections between biological communities (*Biozönosen*) and their environment in a particular section of a landscape (according to Leser, 1980, p. 53). Troll did not consider this to be systematic branch of geography but rather as an approach to an integrating study of landscapes, for which he introduced and developed the art of aerial photography interpretation. This work placed biological and climatological conditions in a central position in German landscape research, the role of geomorphology hence being relatively reduced. Troll considered that landscape ecology was a unifying approach to the natural science aspect of geography, with **social geography** playing a similar role on the human geography side. Uhlig (1973; see also Figure 1.5) similarly used social geography as a term for the integrated approach to geography, but employed a new concept, geoecology to represent Troll's landscape ecology. Uhlig and other German geographers reserve the term 'landscape ecology' for the study of humanity and nature at a higher integrative level, but even then this takes its inspiration mainly from the natural sciences, and particularly from ecology.

While in German the term 'landscape ecology' is fairly clearly defined, in the Anglo-American world it is often confused with such terms as **human ecology** and **ecological analysis**. In his textbook *Geography: A Modern Synthesis* (1983), Haggett proposes ecological analysis as one of the three approaches to integrated geography (Figure 1.6, p. 12). The Anglo-American terms 'human ecology' and 'ecological analysis' put a stronger emphasis on human geography than the German term 'landscape ecology'. This came about because the school of **urban ecology**, which was founded by the Chicago sociologists in the 1920s, concerned itself with the social and functional development and division in urban areas. In such a broad context, 'human ecology' and 'ecological analysis' could be considered as encompassing the functional approach in regional geography, which was developed to study the relationships between centres and their surrounding areas.

Landscape morphology

As a form of regional geography, **landscape morphology** was particularly well developed in Germany between the wars. Otto Schlüter (who played a central role in its development) asserted as early as 1906 that geographers should consider as their unifying theme the form and spatial structure created by visible phenomena on the earth's surface. From this point of view, mountains, rivers, pastures, forests, roads, canals, gardens, fields, villages and towns form a unity in the geographer's eyes: this visible picture is the geographer's object of study. Schlüter (1920), however, maintains that those who try to include too much in their field of vision cannot hope to cope with it all and he therefore argues against the all-encompassing **chorological geography** advocated by Hettner.

Schlüter regarded non-material geographical patterns (such as economic, racial, psychological and political conditions) as not being of primary geographical interest. These should be studied only as part of the explanation of material distributions. The French scholar Jean Brunhes illustrated this when he said that we should study the earth as if we were sitting in a hot-air balloon, looking down upon it. We should analyse the landscape and the characteristic interplay of observable phenomena from up there. The visible landscape is a result of both natural conditions and forces and a manifestation of humanity's work. The landscape itself creates a synthesis; there is no longer a gap between physical geography and the geography of humanity. Both geographies have the same object – the visible landscape – and there is also close contact between them in terms of methods. Because of this, says Leo Waibel (1933), the landscape morphological approach represents an advance for geography, even though the field of inquiry is thereby greatly narrowed.

Landscape morphologists have argued as to how much their field of inquiry should be limited. Should they consider, for example, travelling patterns and the transport of goods and services? Many landscape morphologists considered that these activities should not be objects of their study. They should only be considered as explanatory factors in so far as they contribute to an understanding of the landscape's evolution and character.

Alfred Hettner opposed Schlüter's limitation of geographical study to the visible landscape. He was (as described by Dickinson, 1969, p. 132): 'concerned with the uniqueness of areas, whether this uniqueness was evident in the visible landscape or not. He recognized the focal interest of landscape, but refused to recognize the limits set by it on the study of the human facts in space.'

Hettner maintained the universalistic character of geography on the basis of his evaluations of the philosophy of science. He considered landscapes to be very small *Länder* and, for this reason, regarded attempts to study only their visible parts as a restriction of the scope of regional geography. He argued (1929)

> that it is against the logic of science to permit the picture, i.e. the appearances, to be the decisive factor in defining the subject . . . They exclude thereby exactly those features which have been the pre-eminent concern of geography in the past (human life, nationality, state, partially even the economy) and . . . still are.

Hettner presented his *Länderkundliches schema* as a guideline for a chorological regional presentation, starting with physical features and concluding with population. The presentation of thematic maps (juxtaposed 'on top of each other') makes some sort of integration possible, but it is quite clear that Hettner did not solve convincingly the problem of regional synthesis. Through his follower, Richard Hartshorne, Hettner's dominance of the geographical scene was perhaps greater abroad than at home in Germany. In Germany, the different types of landscape geography did not supersede regional geography – as Hettner had feared – but developed as a bridge between systematic and regional geography. Bobek and Schmithüsen (1949) suggested that landscape could be seen as the integrative product of the **geofactors** of the systematic branches (Figure 1.5, p. 10). Through this integration, only such features which appear in an orderly and regular fashion will be taken into account. Schmithüsen and others have developed methods to build up broader regional types through the delimitation of small units called *Fliese*, or 'mosaic tiles', of landscape (Uhlig, 1967 – see Figure 2.2, p. 53).

The methods employed in landscape morphology were widely adhered to in Europe and, in the USA, Carl Sauer provided a basic evaluation of the new direction in his 'Morphology of landscape' (1925), which was of great importance in the early development of his highly influential **Berkeley School** of geography (see Box 2.9). In 1939 Robert E. Dickinson described landscape geography as the most important new line of growth within the subject, but landscape morphology was, however, developed only by a limited number of British geographers even

Box 2.9   Carl Sauer and the 'Berkeley School'

Throughout his lifetime, Carl Sauer (1889–1975) dominated North American cultural geography in a way that virtually made '**cultural geography**' synonymous with 'human geography' (Jackson, 1989, p. 10). He was appointed head of the Berkeley School of Geography at the age of 33 and held this position until three years before his retirement in 1957. During this time he supervised some 40 PhD students, of whom a great many were later to take up senior academic posts. Through these former students Sauer's influence was assured on a second generation of geographers.

Sauer argued that the best training a geographer could receive came through **field-work** and through developing the skills of observation. A majority of his PhD students studied Latin American and Caribbean topics, and he imprinted on his students 'the need for first hand field experience and for learning the language of the people being studied' (Jackson, 1989, p. 10).

The research Sauer promoted focused on the material aspects of culture, as expressed in the 'cultural landscape', which is most clearly conveyed in his methodological paper 'The morphology of landscape' (1925). In this paper he defined 'landscape' as 'the unit concept of geography', 'a peculiar geographic association of facts'. The landscape was perceived as describing 'a strictly geographic way of thinking of culture' (*ibid.*, p. 30): 'Culture is the agent, the natural area is the medium, the cultural landscape the result' (*ibid.*, p. 46). For example, the USA–Mexico border runs through mountains, deserts and grasslands that are the same on both sides of the border. However, the cultural landscape is different as a result of differences between Mexican and US cultures.

In his early academic years Sauer focused on the morphology of the landscape – as inspired by the German and French regional and landscape studies of static societies that were shaped by centuries of traditional adaptations between humanity and nature. Having observed the dynamics at play in the cultural landscape transitions in America, Sauer shifted his primary research focus to 'a more active appreciation of the social transformation of landscape' (Jackson, 1989, p. 15). His analyses of human agency focused on a search for the origins of institutions and culture traits – for instance, the origin and spatial diffusion of agricultural techniques or the domestication of certain crops or animals. Concentrating on the material aspects of culture, he carved out a special field of study for cultural geography that linked humanity and nature (human and physical geography) and that was not contested by other academic disciplines.

In the tradition of the Berkeley School cultural (and thus human) geography should not concern itself with individuals but with human institutions or cultures: 'The cultural geographer is not concerned with explaining the inner workings of culture or with describing fully patterns of human behaviour, even where they affect the land, but rather with assessing the technical potential of human communities for using and modifying their habitats' (Wagner and Mikesell, 1962, p. 5).

This restriction has, in recent decades, been criticized strongly, from both a humanist and radical point of view. It 'fails to address the wider social context in which cultures are constituted and expressed' (Jackson, 1989, p. 18). New approaches to cultural geography are discussed in Chapters 4 and 5.

Figure 2.2 An example of cartographic techniques in landscape morphology, characterically used by German geographers to build up broader landscape types from a mosaic of local types
Source: After Paffen, 1955

though it was an element in several courses in cartography and was an important component in the interpretation of aerial photographs.

REGIONAL STUDIES IN BRITAIN

'There is fully as much confusion over the meaning of the words "regional studies" as there is over the German word *Landschaft*,' states James (1972, p. 267), with special reference to British geography. **Regional studies** have at least three different connotations in Britain. First, there are regional studies that amount to descriptions of segments of the earth's surface, broadly synonymous with regional geography or *Länderkunde* as defined above. Secondly, there are regional studies that seek to divide the earth's surface into either homogeneous or functional areas of varying size, which we have termed **regionalization**. Thirdly, regional studies may denote regional specialization, when an individual geographer devotes a large part of his or her life to studying different aspects of some part of the world (a good example is Percy M. Roxby of Liverpool, who built up a formidable expertise on China).

Patrick Geddes

In Britain, regional studies were indebted to Vidal de la Blache and other French regional geographers and, to an even greater extent, to the French sociologist Frédéric Le Play (1806–82) through his influential Scottish follower Patrick

Geddes (1854–1932). As a starting point for the study of social phenomena in different parts of the world, Le Play carried out research on **family lifestyles** and family budgets. He recognized that family life depended on the means of obtaining subsistence (that is, **work**), while the character of work is largely determined by the nature of the environment (that is, **place**). This leads to the basic Le Play formula – **place, work, family** – which Geddes transformed into the slogan **place, work, folk** as basic concepts in the study of cities and regions.

Although not a geographer himself, Geddes had a major influence on British geography, especially in the fields of regional survey, regionalization and applied geography. **Field study** – observation and recording in the field – was basic to Geddes' teaching. This led on to what he called '**regional survey**', embracing place, work and folk (alternatively described as geography, economics and anthropology, or as environment, function and organism – Dickinson, 1969, p. 204). Geddes and his followers established the Le Play Society (which grew out of the Sociological Society), to foster regional surveys. Geddes saw an important application of regional survey in **regional planning**. He compared the regional surveyor to an old family doctor who could interpret specialist knowledge and apply it to the actual condition of the individual patient, whom the doctor would know intimately. As a man of radical views, Geddes was rather impatient with geographers who defined their subject as a **descriptive science** – which tells us what *is*. He thought that geography should be an **applied science** that tells us what *ought to be* (Stevenson, 1978, p. 57).

Geddes also influenced the study of **regionalization**. Andrew J. Herbertson (1865–1915), who was an assistant to Geddes in Dundee and later taught geography at Oxford, presented a scheme for a division of the world into **natural regions**, based on an association of surface features, climate and vegetation. The main translator of Geddes' ideas into geography was Herbert J. Fleure (1877–1969) who, from his chair at the University College of Wales, Aberystwyth, rendered them accessible and acceptable to geography. In his paper on 'Human regions' (1919) Fleure, inspired by Vidal de la Blache, sought to reconceptualize the region by taking into account living experience – including the shifting relationships between people. A crucial role in Fleure's accounts of key moments in human history was played by contact zones between different cultures. Like Herbertson, Fleure also worked on the definition of global-scale regions.

A direct application of Geddes' ideas occurs in the work of Charles B. Fawcett (1883–1952) who translated the somewhat obscure ideas of Geddes into a workable form and made one of the first identifications of **functional regions**. In his book *The Provinces of England* (1919), which attracted considerable attention, he proposed for England a federal structure of 12 provinces. The six principles of the division were clearly inspired by Geddesian regionalism, particularly so in the organization of each province round a 'definite capital which should be the real focus of regional life', the drawing of provincial boundaries 'near watersheds, rather than across valleys', and paying 'regard to local patriotism and tradition in the grouping of areas' (Stevenson, 1978, p. 59).

THE DEVELOPMENT OF APPLIED GEOGRAPHY

One concept derived from Geddes' work was that of the **regional survey** of potential land quality and land use as a basic input to plans for economic development. During the 1930s and on his own initiative, L. Dudley Stamp (1898–1966) organized and directed the first **British Land Utilization Survey**, employing some 22,000 schoolchildren in the mapping of **land use** on a scale of 1/2,500 in their home district under the supervision of school and university teachers. Eventually, the first Land Use Survey formed the basis for an official agricultural regionalization of Britain. When the Second World War began in 1939, the vital importance of the maps was quickly appreciated as essential data for the programme of food production expansion necessitated by the German blockade. At the International Geographical Congress in Lisbon (1949), Stamp suggested establishing a **World Land Use Survey**. The idea was adopted and an international commission under the IGU (the International Geographical Union) was set up to supervise the work.

In the 1930s, economic depression accentuated the effects of longer-term economic changes that were creating marked divergences of economic prosperity and distress between British regions. Geographers contributed to national and local studies of these prosperous and 'depressed' areas – although description predominated analysis at this stage. During the 1940s, geographers were widely involved in planning the postwar reconstruction. And during the 1980s, several geographers claimed to have recognized a re-emergence of marked variations in prosperity between the regions of Britain. This latter work, which was obviously politically controversial, showed considerable advances in analytical techniques from that conducted earlier in the century.

In recent years there has been an increasing interest in the **cultural landscape** in connection with **regional planning** – perhaps as a reaction against the uniformity of most modern buildings and other landscape features. The ease with which a landscape may be disfigured through technology has awakened a concern to care for the landscape. Uncontrolled building has created so many eyesores that the need has arisen for an active policy of landscape protection – including both aesthetic and economic aspects. This had already been pointed out by Steers in his book *The Coastline of England and Wales* (1946), which gave a clear indication of the major planning problems of coasts – including the notorious bungalow and campsite development which had been permitted during the interwar years (Freeman, 1980, p. 53).

J.H. Appleton (1975) and others have demonstrated the importance of an informed view of landscape if the attractiveness of long-settled countries like lowland Britain is to be preserved. The rapid development of industrial archaeology (especially as a hobby of educated laypeople in Europe) also illustrates a persistent general belief in the importance of preserving something of the past's visual culture. Robert Newcomb (1979) calls for a 'planning of the past' and gives suggestions for the active use of historic features so that the preservation of at least relics of historic landscapes and townscapes is possible. Important recent texts on the issues surrounding landscape heritage and protection are David Lowenthal's *The Past is a Foreign Country* (1985) and *Possessed by the Past. The Heritage Crusade and the Spoils of History* (1996).

Demand for area planning in districts, counties and regions has similarly increased interest in studies of the cultural landscape. In Norway, for instance, the book *Mountain Regions and Recreation* (*Fjellbygd og feriefjell*) (Sömme, 1965) – the result of research into the effects of the recreational use of upland areas – was a landscape study that attempted to provide a sound basis for new planning laws.

A newer trend, connected to *humanism* and *postmodernism* (see Chapter 4 and 5), are work on the interpretation of landscape as text. Important contributions are for instance Denis Cosgrove's (1984) book *Social Formation and Symbolic Landscape* and Kenneth Olwig's (1984) *Nature's Ideological Landscape*. Another important text is *The Iconography of Landscape* edited by Denis Cosgrove and Stephen Daniels (1988).

Several **regionalization** projects have sought to delimit single-attribute regions – such as industrial, climatic, vegetational, morphological and social. While each project was linked with the relevant systematic science – social regions with sociology, for example – geographers focused on the region, choosing either the **single-attribute region** of the specialist or the **multi-attribute region** of the synthesizer (Johnston, 1997, p. 47).

Whereas between the wars much effort was put into the development of methods to define multi-attribute regions, after the Second World War systematic studies in various specialisms gradually claimed the attention of the great majority of British geographers. Even those who regard all forms of regional geography as blind alleys concede they will continue to exist within geography. E. A. Wrigley (1965, p. 13) put it this way:

> The regional period of geographic methodology like the 'classical' (including determinism) has left many traces, some of which will perhaps prove permanent, on the methods used in organizing and presenting geographical material. Any discipline is both the product and the victim of its own past successes and these were two of the most important successes thrown up by geographical scholarship.

In the past, regional geography's widespread hold discouraged many geographers from seeking general relationships and theories, and led them to decry the formulation of geographical laws and models. Rejecting the general theories of the determinists, they sought refuge in regional methodologies where each area is unique and somewhat exceptional and must be studied as such. They upheld that geographers should continue with **idiographic** methods – that is, the description of unique phenomena and unique regions.

Adherents of the regional school regard geography to be as much an art as it is a science, and consider that geographers (in common with other 'arts' scholars) should emphasize literary quality. Jan Broek (1965, p. 21) argued that 'The humanities stress real persons and cases rather than models, quality rather than quantity, evaluation and evocation rather than calculation, beauty and wisdom rather than information. Geography shares these attitudes to some extent.'

In the 1950s and 1960s the idiographic method and the humanistic way of thinking within the subject were strongly criticized. A new and dynamic school developed and new methods brought into use. There was much talk of paradigm crises and revolution. We shall see what this implies in Chapter 3, which attempts to bring together the threads from this chapter and to put the 'quantitative revolution' into its historical context.

3

Paradigms and Revolutions

Geography is concerned to provide accurate, orderly, and rational description and interpretation of the variable character of the earth's surface.

(Hartshorne, 1959, p. 21)

A traditionally held view – that geography is concerned with giving man an orderly
description of his world – makes clear the challenge faced by contemporary geographers
. . . The contemporary stress is on geography as the study of spatial organisation expressed as patterns and processes.

(Taaffe, 1970, pp. 5–6)

Geography can be regarded as a science concerned with the rational development, and
testing, of theories that explain and predict the spatial distribution and location of
various characteristics on the surface of the earth.

(Yeates, 1968, p. 1)

While geographers would be in general agreement that major changes took place
in every aspect of geographical thought during the 1950s and 1960s, there is no
general consensus as to what degree the many innovations that took place during
those decades will be of lasting value, nor on their effect on the future of geography as an academic discipline. The quotations above illustrate the divergent
views expressed during this period.

Hartshorne and Yeates seem to differ widely in their prescription of methods
but find more common ground in their definitions of the objectives of geographical research. While both are concerned with the variation of phenomena over the
earth's surface, Hartshorne visualizes geography as an **idiographic science**, with
its main emphasis on the description and elucidation of individual phenomena
because these are unique. Yeates (*ibid.*, pp. 3–7) considers geography to be a
nomothetic (law-giving) science that requires the development and testing of
theories and models through **hypothetic-deductive methods** in order to develop
geographical laws.

Taaffe suggests that geography had changed by 1970 – or was in the process of
changing from an idiographic to a nomothetic science. Explanatory models that
were once thought to be satisfactory for geographical work were discredited by a
large number of geographers. John D. Adams (1968, p. 6) said that

geography is currently in the throes of a paradigm crisis. Instead of asking the traditional question 'Is it geography?', or 'What *is* geography?', geographers are now asking 'What should geography be?' If a satisfactory answer is not found to the latter question the next question is likely to be 'Is geography relevant?'

It is clear from Chapter 2 that this was not the first crisis in geography's development: Ritter's teleological framework did not satisfy the determinists; and views that were scientifically acceptable to Ratzel and Semple were too deterministic for Vidal and Hartshorne.

KUHN'S PARADIGMS

Perhaps Thomas S. Kuhn (1962/1970a) was right when he claimed that science is not a well regulated activity where each generation automatically builds upon the results achieved by earlier workers. Instead it is a process of varying tension in which tranquil periods, characterized by a steady accretion of knowledge, are separated by crises. These crises can lead to upheavals within disciplines and breaks in continuity. Even Hettner (1930, p. 356) had suggested

> that a science does not always follow a straight line of development, but often zig-zags on its road to higher professionalization. This development often corresponds to changes in generations among its professional practitioners. It would be a bad thing if a new generation had no new thoughts . . . [but] a science should not suddenly be changed into something quite different from what it has been before.

In his last remark, Hettner was upholding the importance of historical continuity.

Kuhn rejected such veneration of the past, arguing instead that fundamental changes are often necessary to enable science to progress. While it is possible to determine objectively whether an explanatory framework is satisfactory and reasonable *within* a specific scientific tradition, we have to choose between different scientific traditions – and this choice is subjective. We must select what Kuhn calls **paradigms** (models or exemplars) for our science.

Kuhn defined paradigms (1962/1970a, p. viii) as 'universally recognised scientific achievements that for some time provide model problems and solutions to a community of practitioners'. Haggett (1983, p. 21) defines a paradigm as 'a kind of supermodel. It provides intuitive or inductive rules about the kinds of phenomena scientists should investigate and the best methods of investigation.' A paradigm is a theory of scientific tasks and methods that regulates the research (for example) of most geographers or, where there is a conflict between paradigms, of a group of geographers. The paradigm informs researchers what the object of their science should be, to which questions they should try to find 'acceptable' answers and which methods can be considered as 'geographical'.

Initially, Kuhn did not provide an altogether clear definition of the concept of paradigm. As Mair (1986) has pointed out, Kuhn later clarified some ambiguities and accepted that he had conflated two conceptually distinct, though empirically inseparable, types of paradigm. In the second edition of his book, Kuhn (1970a) argued that the most basic function of a paradigm is as an **exemplar**: a concrete problem solution within a discipline that serves as a model for successive scientists. Generally, such exemplars tie a scientific theory together, serving as an example of a successful and striking application. The most important paradigms are those that generate whole new fields of scientific endeavour. Paradigms (in the sense of 'exemplars') may not always have this all-embracing effect, but will guide research as they are presented to students as models the students should try to copy. An example from geography would be the **regional monographs** written by Vidal de la Blache and some of his contemporaries, which established examples that have served a long-standing geographical tradition.

The other meaning of paradigm put forward by Kuhn is as a **disciplinary matrix** – 'the entire constellation of beliefs, values, techniques and so on shared by the members of a given community' (1970a, p. 175). A disciplinary matrix may be shared by a large group of members of a discipline while, at the same time, each member is working with different 'exemplars' in his or her everyday research (Mair, 1986, p. 352). It is in the sense of disciplinary matrix that the term paradigm has most commonly been applied to geography. The term 'paradigm' is used here in its disciplinary matrix sense as we attempt to interpret the history of geography in (simplified) Kuhnian terms. According to this Kuhnian model, scientific development consists of a series of phases (Figure 3.1; Box 3.1).

Box 3.1   Kuhn's theory of the development of a science

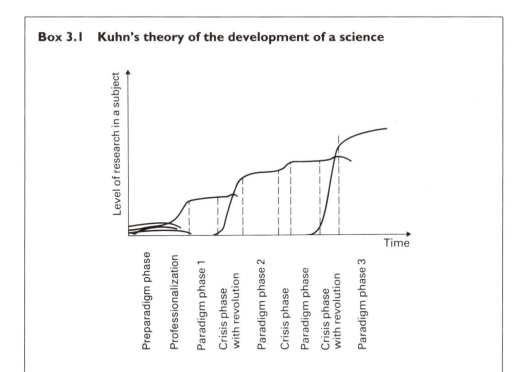

Figure 3.1 A graphical interpretation of Kuhn's model of a discipline's development. This figure indicates that paradigms and paradigm shifts dominate the whole discipline, whereas in his later work Kuhn pointed out that many scientific revolutions only impinge on the work of smaller groups of scientists within a discipline
Source: After Henriksen, 1973

In a branch of science that becomes the subject of thorough and systematic study, the **preparadigm period** is marked by conflicts between several distinct schools which centre around individual scientists. If we take paradigms to mean **exemplars** (that is, concrete examples of research to be followed), this phase could be labelled a multi-paradigmatic period, as each school of thought develops its own model solutions. **Professionalization** (or the subject's development from the preparadigm period to the stage of scientific maturity) has taken place at different dates amongst the different sciences. Kuhn argues that mathematics and astronomy left the preparadigm phase in antiquity. The transition begins when the question as to what a specific science is about becomes acute. This happened in geography when a university degree in the discipline was needed to qualify as a high-school geography teacher. A disciplinary matrix had to be defined for such a degree course, which would also secure and demarcate geography's domain from other university disciplines. One of the conflicting schools of thought will often dominate the others in that it seems best suited to win the discipline academic esteem. A paradigm is established that leads to concentrated research within a clearly distinguishable problem area – an activity described as **normal science**.

To bring a normal research problem to a conclusion is to achieve anticipated results from new empirical sources; Kuhn called this '**puzzle-solving**' because of its similarity to solving jigsaw or crossword puzzles. The research workers' perception is constrained by their paradigm; their observation of data is hence directed or attracted to the expected result; there is in-built opposition to unexpected discoveries. The paradigm can advance research and provide economy in the amount of research needed: research workers can go straight to the research frontier without having to define their philosophical bases and underlying concepts. Group identity among research workers is also a strong psychological advantage, which itself can stimulate scientific productivity.

Sooner or later a period of '**normal science**' is replaced by a **crisis phase**. This happens when more problems are accumulated than can be solved within the framework of the ruling paradigm. The crisis phase is characterized by a reassessment of former observational data, new theoretical thinking and free speculation. This phase involves basic philosophical debates and a thoroughgoing discussion of methodological questions. The crisis phase may end when it seems that no significantly better theory can be developed to solve the problems and thus, consequently, research must continue using the old paradigm. Alternatively, the crisis phase ends when a new paradigm attracts a growing number of researchers away from the old paradigm.

The acceptance of a **new disciplinary matrix** inaugurates a **revolutionary phase**. This means a break in the continuity of research and a thoroughgoing reconstruction of the research field's theoretical structure, rather than steady development and the accumulation of knowledge. Accepting a new paradigm is also revolutionary because it attracts the allegiance of the younger research workers who are opposed to established scientists. The new scientific 'reason' seldom triumphs by convincing its opponents; rather, it succeeds as they die and a new generation takes over. Younger workers who do not conform to the newly accepted paradigm are ignored by its followers, and researchers are continually forced to ask themselves whether the type of 'puzzle-solving' they are doing is the 'right' one.

Exchanging one paradigm for another is not a wholly rational process. The new paradigm will generally provide solutions for the problems the old one found difficult to resolve, but may not answer all the questions that were fairly easy to solve before. It is also seldom possible to argue logically that the new paradigm is better than the old. Even if a new paradigm can buttress itself with empirical and logical proofs, its original choice was basically subjective – an act of faith. Aesthetic considerations may influence the choice of a new paradigm – it may be regarded as simpler or more beautiful than the old one.

To many, Kuhn's picture of scientific activity is alarming: our faith in the objectivity of research is weakened when we consider how subjective the choice of paradigms can be and when we experience the often protracted opposition of some scientific workers to the establishment of new explanatory models. Few research workers welcome a general debate on the subjectivity of research – it may lead to the evaporation of respect for the research and the loss of financial support. On the other hand, as suggested by Peter Taylor (1976, p. 132), the youngest research workers who are at the bottom of the formal academic hierarchy have a clear vested interest in changing the existing scientific ideology, and thereby taking over from their elders.

The Kuhnian model has given the 'new prophet' a very effective weapon against the disciplinary matrix of a scientific 'establishment'. They do not need to justify their research as objective in itself; it is enough if they declare it to be objective within the subjective framework they have chosen. This can cause conflict amongst social scientists, as it is all too easy to equate the choice of a paradigm with the adoption of a particular value judgement. The ultimate conclusion may be that only those who affirm the same general outlook on the world and who have similar political beliefs are competent to evaluate a piece of scientific research.

CRITICS OF KUHN

Few geographers, said Bird (1977), have noticed that the concept of a paradigm ruling a community is akin to that of a dogma – which must be adhered to if a person is to be accounted orthodox. This is in line with the views of Karl Popper, the scientific methodologist who has most effectively criticized Kuhn. Popper (1970) maintained that an active and progressive science should be in a constant state of revolution. While acknowledging that Kuhn had demonstrated the existence of 'normal science' and 'paradigms', Popper deplored these periods as dangers to scientific progress. 'Normal science' becomes established when uncritical scientists accept the leading dogmas of their day and espouse some newly fashionable and formerly revolutionary theory. Popper feared that if scientists of this type were to dominate scientific thinking, this would herald the end of science. Dubbing 'normal science' the 'myth of the framework', Popper asserted that one of the scientist's most important roles is to break down myths. Given that we are always trapped within theoretical frameworks to some extent, we can break out of them at any moment if we act as true scientists. An active science will be in a state of **permanent revolution** (Bird, 1975). Some scientists have interpreted Popper's concept of a 'permanent revolution' as a prescription for what science should be rather than as a description of how it is actually practised. Kuhn (1970b), however, declared that Popper's demand for a 'permanent revolution' was built on assumptions that are just as unreal as are attempts to square the circle.

Paul Feyerabend rejected as historically false the Kuhnian model of alternating periods of 'normal science' and 'revolution'. He maintained that even in theoretical physics, which Kuhn (1962/1970a) used as his example, there have always been alternative basic theories that could act as 'exemplars'. Sciences do not show a chronological shift between periods of normal science and periods of pluralism. Hence, the historical development of a science might best be described by synthesizing the models of Popper and Kuhn.

This outline of Kuhn's ideas has largely been formulated within the context of a 'natural science', and a natural science does not question the basic implicit worldview that science is the study of the empirical world. A fundamentally different approach to the interpretation of the history of science was presented by the social theorist Michel Foucault (1972; 1980). Foucault's concern was to examine the political status of science and the ideological functions it could serve. Central to his argument is that there is a fundamental connection between power, knowledge and truth. Each society has its regime of truth, its 'general politics' of truth, i.e. the types of discourse it accepts and makes to function as true (Foucault, 1980, p. 131). Truth is therefore a relative concept, depending on the power relations within the society that produces it. To interpret changes in science throughout the course of modern history, Foucault focuses on the various worldviews (or structures of thought) people have held. These he calls **epistemes**. When epistemes change then science will change as well. This implies that societies create relative truth in order to reinforce the power relations within that society. We will return to this value discussion in Chapter 4. The discussion between Kuhn, Popper, Feyerabend and Foucault depends on their different ideas as to how a scientist works and how scientific theories and laws are established. At this point we therefore need to look briefly at models of scientific explanation.

INDUCTION, DEDUCTION AND ABDUCTION

Francis Bacon (1561–1626) defined the **inductive** route to scientific explanation (Figure 3.2). A scientist starts with a range of sense perceptions he or she works up conceptually and verbally into a number of loosely arranged concepts and descriptions that we like to call facts. Next, certain definitions are necessary to organize the data. Afterwards the facts are evaluated and arranged in relation to the definitions.

The ordering and classification of data are often the chief activity of a science in the early stages of its development. These first classifications may have only a weak explanatory function. Continuing study of the interaction between classes and groups of phenomena reveals a number of regularities; such regularities and laws may be called **inductive laws** since they are derived from the observations of a large number of single instances. The inductive route's weakness is, of course, the jump from a number of single instances to a general truth ('All swans I have seen are white: all swans are white').

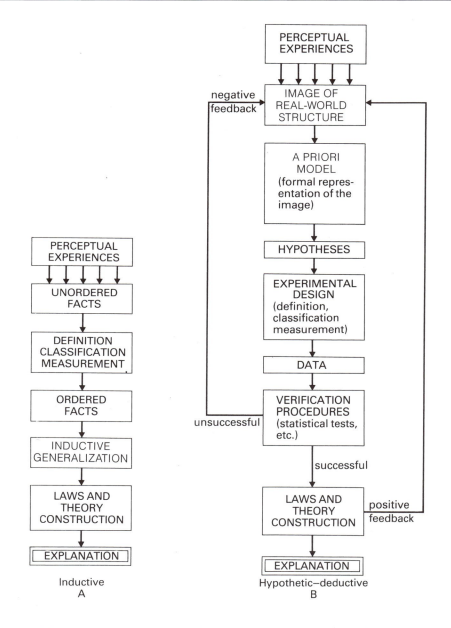

Figure 3.2 Inductive and hypothetic-deductive routes to scientific explanation
Source: From Harvey, 1969

Here we must clarify what a **scientific law** is. Braithwaite (1953, p. 12) defined a law as 'a generalization of unrestricted range in time and space', in other words, a generalization with universal validity. With this definition we can distinguish between **empirical generalizations** and **laws**. An empirical generalization is valid for a specific time and place but a law is universal. James (1972, p. 473) maintained that a law within Braithwaite's rigorous definition can hardly be formulated on the basis of geographical evidence. The only truly universal laws are those of physics and chemistry, although even in physics there are elements of uncertainty that make probability calculations necessary. Harvey (1969, p. 31) gave the concept of law a much wider significance and postulated a threefold hierarchy of scientific statements from **factual statements** or systematized descriptions, through a middle tier of **empirical generalizations** or laws, to **universal theoretical laws**.

Since the nineteenth century inductive arguments have been increasingly replaced by **hypothetic-deductive methods** (Figure 3.2b). Research workers, starting from an inductive ordering of their observations or from intuitive insights, try to devise for themselves **a priori models** of the structure of reality. These are used to postulate a set of **hypotheses** that may be confirmed, modified or rejected by testing them with experiments using empirical data. A large number of confirmations are supposed to lead to the **verification** of a hypothesis, which is then, for the time being, established as a law and basis for theory construction.

Karl Popper pointed out, however, that the truth of a law does not depend on the number of times it is confirmed experimentally; it is easy enough to find empirical support for almost any theory. The criteria for its **scientific validity** are not the confirmatory evidence, but that those circumstances which may lead to the rejection of the theory are identified. It follows that a theory is scientific if it is possible to **falsify**. Kuhn criticized Popper for believing that a theory, or hypothesis, would be abandoned as soon as evidence is found which does not fit it. Kuhn maintains that all theories will eventually be confronted with some data which do not fit. A fundamental theory is not rejected if individual research data do not fit it, for if it were, then all theories would have to be rejected. Up to now the history of science does not record any theory that has not eventually been confronted by contradictory circumstances (or instances of falsification). According to Kuhn, a fundamental theory is only rejected when a new theory is put forward that is *believed* to be superior (Johansson, 1973).

Feyerabend added that scientific development is much more **irrational** than Popper's scheme of falsification allows. As well as straightforward arguments, the proponents of new theories have also often used propaganda and psychological tricks. Feyerabend maintains that the development of scientific knowledge follows an irrational, almost anarchic path, along which almost anything goes, as far as methodology is concerned (Åquist, 1981, p. 11).

Kuhn did not accept Feyerabend's views on the irrationalities of scientific progress. He denied that he intended to present scientific theories as intuitive and mystical, more appropriate for psychological analysis than for logical and methodological codification. Kuhn asserted that every scientist must gather as much rational proof in support of a new theory as possible and to be precise and honest in his or her work. On the other hand, we need to realize that *data are dependent on theory*. All observations presuppose a certain conceptual apparatus by which the sense perceptions, or empirical verifications, can be arranged. Expectations and former ideas, that is, the hypotheses which have been set up, guide the interpretation of the data to a large extent.

Bird (1993, p. 2) points out that there is no agreed description of *the* scientific method, and continues:

(a)

(b)

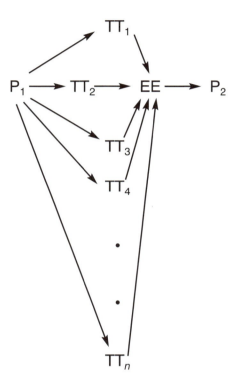

Figure 3.3 The hypothetic-deductive method as presented by Popper: P_1: initial problem; TT, TT_1–TT_n: tentative theories; EE: error elimination; P_2: new set of problems. (a) Sequence of scientific work with testing of one tentative theory; (b) sequence of scientific work with scientific examination of several competing tentative theories
Source: From Popper, 1972, pp. 119, 243; Haines-Young and Petch, 1986

Just imagine the situation if it were. A totalitarian world of procedures would have to be learnt and obeyed. It is difficult to imagine such a universal framework lasting for very long. But that is not to say that there are no rules or tricks of the trade, although such as now seem useful are all on probation.

Bird (*ibid.*; in relation to human geography) and Haines-Young and Petch (1986; in relation to physical geography) find Popper's version of the hypothetic–deductive method and his **critical rationalism** most useful for scientific inquiry. The critical rationalist view of science can be said to build on three principles:

1) *The principle of falsification:* universal statements and theories can only be re-futed, not verified.

Figure 3.4   Critical rationalism
Source: Haines-Young and Petch, 1986, p. 45

2) *The principle of criticism:* scientific knowledge grows only when open to critics, trial and error.
3) *The principle of demarcation:* the characteristic of scientific statements are that they are empirically testable, capable of refutation if they are false. (*ibid.*, p.44).

The starting point for any scientific project (see Figure 3.3) is **problem formulation**. This is often a difficult intellectual exercise. Having discovered a problem (P_1 in Figure 3.3) and formulated the restricted questions within the problem area you might be able to analyse in your research, the next step is to decide on how to proceed and then to follow the chosen procedures. In practice this means that, after the problem formulation, you will look at earlier research within your problem area and then discuss and evaluate tentative **theories** (TT) relevant to your research questions. Then you need to find relevant methods for empirical testing, or **error elimination** (EE) of the theories. This may lead on to a residual irreducible problem or problems (P_2), which may be different from the initial problem. P_2 may lead on to suggestions for further research.

In a scientific thesis it may only be possible to test one tentative theory (Figure 3.3(a)), but this often means a restricted choice between several hypotheses or tentative theories. Figure 3.3(b) presents an alternative procedure as prescribed by Popper (1972, p. 243):

> An excellent example of the method in operation is provided by the work of Battarbee *et al.* (1985) on the causes of lake acidification in Galloway, Scotland. They examine the hypothesis that acidification is due to acid precipitation against competing hypotheses that it is due to heathland regeneration, afforestation or long-term post-glacial natural acidification.

(Haines-Young and Petch, 1986, p. 63

Figure 3.5 Deduction, induction and abduction. On abduction the first arrow from empirical data to empirical rules and laws is 'dashed' to indicate that the real abduction starts with the 'lift' from empirical patterns to tentative theory
Source: From Alvesson and Sköldberg, 1994, p. 45.

Popper maintains that it is only through criticism that false ideas can be eliminated and theories gradually improved. Since the truth of an observation can never be established finally, neither can our theories. The best we can say is that our evidence **corroborates** (makes more certain) a theory (Figure 3.4).

Popper's **hypothetic-deductive method** is based on deduction from theories that have been created in the imaginative mind. Alvesson and Sköldberg (1994, p. 43) maintain that such imaginatively inspired theory seems to be a form of scientific virgin birth – probably as rare in science as in nature. Tentative theories, or hypotheses, are almost always based on previous reflections on facts – an inductive phase. **Induction** starts with facts and deduction with theory, but it is not necessary to choose one of them exclusively. **Abduction** is the third alternative, recommended by Alvesson and Sköldberg (*ibid.*).

Abduction, like induction, is based in empirical facts (Figure 3.5), but the real process starts with the 'lift' from empirical patterns to tentative theories. Then follows the deductive process as proscribed by Popper – with critical testing leading on to corroboration or refutation, new problems and new testing, in a never-ending scientific process. Abduction may be likened to the process of medical diagnosis: the doctor observes the patient's symptoms and assumes they are caused by a particular disease. But his or her assumptions must be confirmed by tests and checked against other symptoms or comparison with data from other patients. In a scientific process there will be series of abductive 'jumps' between inductive and deductive reasoning, or between empirically loaded theory and theoretically loaded empiricism.

In this section we have discussed general theories of science that have mainly been developed with reference to theoretical natural sciences, such as physics. The transference of natural science methods into the social sciences has, however, brought about a vigorous debate among social scientists that will be discussed in more detail in Chapter 4. There may be similar problems in transferring the

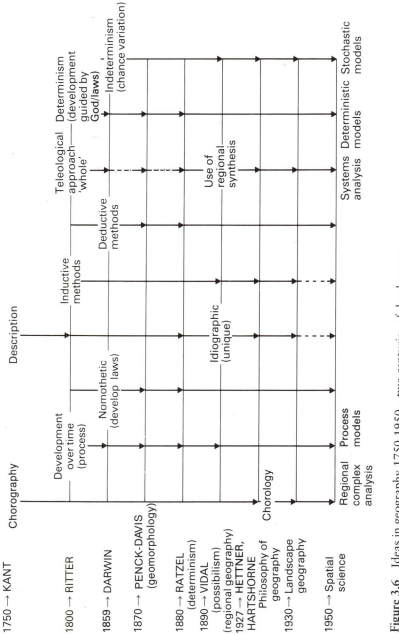

Figure 3.6 Ideas in geography 1750-1950 – two centuries of development

paradigm concept. Kuhn was trained as a physicist, and his theory derives largely from a study of the history of physics. How far is the history of physics relevant to the less theoretical and quantitative sciences? To answer this question we will now look at the history of geography in the light of Kuhn's model.

CHANGING PARADIGMS IN GEOGRAPHY?

Bird (1977) has argued that Kuhn has been the most influential scientific methodologist as far as geography is concerned. Mair (1986) suggests that geographers influenced by Kuhn fall into two groups. First there are those who have used Kuhn to legitimize their propaganda for a 'paradigm change' within the discipline and as a weapon against the scientific 'establishment' (see p. 85). Secondly, several geographical historiographers have tried to apply a Kuhnian model to the development of geographical thought (see, among others, Widberg, 1978; Schültz, 1980; Harvey and Holly, 1981; Stoddart, 1981; Martin, 1985; Johnston, 1997). The paradigm concept has taken on a life of its own beyond that originally envisaged by Kuhn, and as such has been regarded as a useful 'exemplar' (model or teaching framework) for histories of geography.

Figure 3.6 attempts to systematize the theoretical development of geography up to the 1950s, but it gives an incomplete and oversimplified picture as only the main concepts in the subject's development are shown. Over the course of time, these concepts have changed in significance and connotation. In Kuhn's terminology (Figure 3.1), until Darwin's time geography was in its preparadigm phase: Kant did not found a school of geography but indicated a role for the subject and suggested its position in relation to other sciences. Geography was, for him, a chorographic and mainly descriptive science, distinct from the systematic sciences and from history.

Ritter, on the other hand and contrary to Kant and his school, did not emphasize the distinctive roles of geography and history but instead emphasized developments over time, linking history with geography. Ritter was, however, the first geographer to describe his method clearly, and his account conforms to Francis Bacon's classical model of how a scientist works (Figure 3.2(a)). However, because of Ritter's teleological outlook, this first apparently active school of geography did not lead the subject into its first phase of a paradigm: contemporary developments in Darwinism meant the rejection of those ideas that might have led to a paradigm.

It should be emphasized that Darwinism did not represent a complete break with the major ideas upon which Ritter's geography had been founded. The study of development over time was still regarded as very important and a deterministic explanatory framework was strengthened further. The break with Ritter was about the forces that shaped development. After Darwin, scientists looked for the laws which controlled nature (and for materially conditioned social laws) and so, to a considerable extent, they adopted a **nomothetic** (law-making) approach to science.

Neef (1982, p. 241) may be right when he suggests that it was at this time that the most important revolution took place in geography, when the universities subdivided their faculties into separate disciplines and a **cosmographic** way of thinking was replaced by the **causal explanations** characteristic of the developing natural sciences.

During the latter half of the nineteenth century, to make geography acceptable as a science the newly appointed geography professors tried to recast the discipline as a nomothetic science. The hypothetic-deductive method was not, however, applied in a strict sense; we may agree here with Minshull (1970, p. 81) that determinists, as well as geomorphologists, stated generalizations first and then supplied a few highly selected examples as proof. Unlike physicists, geographers could not test hypotheses by verification procedures that involved a number of repeated experiments; statistical tests that might have played the same role as experiments were not as yet sufficiently developed to cope with complex geographical material.

Geomorphology and **determinism** may be said to represent geography's first paradigm phase. This paradigm was effective for geomorphology – it lasted for a good half-century and advanced the whole subject's scientific reputation through its accumulation of scientific information until alternative explanations could be put forward. While geomorphology expanded other branches of geography experienced a series of crisis phases. For this reason we can leave geomorphology to one side for the time being and concentrate on developments in human geography.

As the dominating paradigm in human geography, determinism had a short life, being challenged by the **possibilists** and the French school of regional geographers. These geographers stressed the idea that humanity has free will and participates in the development of each landscape through unique historical processes. Methodologically, geographers were trained to concentrate their study on the unique single region. This inevitably limited the development of theory (as normally understood in science) and made the hypothetic-deductive method redundant. The appropriate methodology under this approach would be to try to *understand* a society and its habitat through **field study** of the ways of life and attitudes of mind of the inhabitants. Such methods (in the form of **participant observation**) characterize the work of many social anthropologists today.

Fieldwork was regarded as of the utmost importance in the French school of regional geography. Vidal de la Blache based his *Tableau de la Géographie de la France* (1903) on studies in each *département*. Albert Demangeon walked every lane in Picardy before publishing his regional monograph on that *pays* in 1905. This fieldwork was, however, mainly concerned with presenting a picture of the material ways of life in the regions and had to be supplemented with the collection of factual material from statistical, historical and archaeological sources. In its handling of data – in the data's organization, classification and analysis – the regional approach to geography resembled the inductive method very closely. Regional geographers also sought general causal relationships but were rather unwilling to identify these as 'laws'. More explicitly **qualitative methods**, such as **participant observation**, were only adopted directly in local studies much later.

Although possibilists reacted against the determinists' simple explanatory models, many of their ideas were derived from Darwinism. They took over Darwin's concepts about struggle and selection although they also considered that chance and human will played an important role in development. While possibilism could be said to constitute a new paradigm, it did not immediately replace determinism. Partly because of the strength of geomorphology and physical geography, the deterministic explanatory model continued to survive side by side with possibilism.

For a long time, however, geographers continued to stress the central position of **regional geography**. Georges Chabot declared in 1950 (p. 137) that 'Regional geography is the centre around which everything converges'. It is, however, fairly obvious that the greatest advances in geographical research during the twentieth century have taken place within **systematic geography**. In geomorphology, bio-geography, economic geography, population geography and many other branches of the subject, a range of new theories and methods have evolved. During the interwar period, landscape ecology and **landscape morphology** were subdivided, and many specialist studies were made on **urban morphology**. Research into the morphology of rural settlements was also separated from general studies of agrarian cultural landscapes.

Regional geography flourished in such countries as France, where in school and university teaching geography was closely associated with history and where the educational system fostered a national self-image of sturdy peasantry and cultured townsfolk. Regional studies were also important to the academic leaders of the emergent nations in central and eastern Europe, who were seeking to establish and preserve the uniqueness of their national heritage. This was to be achieved not just through their native languages but also by studying a whole range of traditional relationships with their lands, which had survived centuries of foreign domination.

While the peace settlement of 1919–21 created many new European nation-states, arguments over boundaries between the 'winners' and 'losers' of the war continued to draw extensively on local historical and geographical relationships. The economic revival of western and northern Europe from prosperity to affluence after 1945 has, however, been associated with the growth of essentially similar urban industries and services organized nationally (but today greatly influenced by the **globalization** of the economy). 'Regions' have come to be defined in strictly economic terms: 'regional policies' are devised to help areas that lag behind current norms of economic growth.

In the USA, Edward A. Ackerman (1911–73) argued that 'taken as a whole, those geographers who had mastered some systematic field before the war were notably more successful in wartime research than those with a regional background only' (1945, p. 129). He went on to point out ways in which emphasis upon systematic methods would best serve geography in the future. Although an immediate impact of his work is difficult to trace, his analysis encouraged the subsequent move towards training on the systematic side (White, 1974, p. 301).

Another reason for the limited progress of regional geography was the basic philosophy for the subject held by Hettner and Hartshorne. While both regarded the **regional geographical synthesis** as central to geography, they discouraged historical methods of analysis, arguing (with reference to Kant) for geography to be regarded as a chorological science. Hartshorne was strongly criticized by, among others, Carl Sauer who, within a year of the publication of *The Nature of Geography* in 1939, said: 'Hartshorne . . . directs his dialectics against historical geography, giving it tolerance only at the outer fringes of the subject . . . Perhaps in future years the period from Barrow's 'Geography as human ecology' (1923) to Hartshorne's latest resume will be remembered as that of the Great Retreat' (Sauer, 1963, p. 352).

The concept of the subject Hettner and Hartshorne had developed was, however, adopted by a large majority of human geographers from the 1930s until the 1960s. This, if anything, could be regarded as a **paradigm**. The disadvantage was that it did not lead to a universally accepted method of chorological regional description. Neither Hettner's *Länderkundliches schema* nor Hartshorne's identification of regions through 'comparison of maps depicting the areal expressions of individual phenomena, or of interrelated phenomena' (1939, p. 462) solved convincingly the methodological problems of regional synthesis. Vidal and the French school of regional geography, on the other hand, produced scientific works that served as **exemplars** for a large group of students, and thus could be said to have functioned better as a paradigm.

Schlüter and his followers also provided a basis for the study of the cultural landscape by developing methods of **landscape morphology** (see pp. 50–53). However, the majority of geographers, including the influential Hettner, regarded these methods as being too restricted in scope for geography as a whole, since landscape morphology restricted its analysis to the visible landscape whereas a proper regional synthesis also includes the 'invisible' transactions of social and economic life.

Kuhn's model, so far, fits the development of geographical science only superficially. As we have followed the early progress of the subject we have seen how new paradigms (in the sense of '**disciplinary matrices**') have, to some extent, included ideas from older paradigms. New paradigms therefore lose clarity and value as a guide for research until, in the end, more and more people define geography as what geographers do. Despite the impressions we may get from simplified accounts (for instance Wrigley, 1965), a closer look at the history of geography reveals that complete revolutions have not taken place; paradigms, or what may be more appropriately termed schools of thought, continue to exist side by side.

AN IDIOGRAPHIC OR NOMOTHETIC SCIENCE?

Another reason why paradigm shifts can be regarded as more apparent than real is that each new generation of workers, or each individual trying to change the scientific tradition of the discipline, will tend to ascribe a more fundamental significance to their own findings and ideas than they really have. A number of times in the history of geography we have witnessed a characteristic oversimplification of the views held by the immediately previous generation or, rather, of those held by the leading personalities of the current tradition.

A rather good example is the vigorous criticism of Hartshorne presented by Fred Schaefer (1953) in 'Exceptionalism in geography'. Schaefer attacked the 'exceptionalist' view of the Kant–Hettner–Hartshorne tradition – the view that geography is quite different from all other sciences, methodologically unique because it studies unique phenomena (regions), and therefore is an **idiographic** rather than a **nomothetic** discipline:

> Hartshorne, like all vigorous thinkers, is quite consistent. With respect to uniqueness he says that 'While this margin is present in every field of science, to greater or lesser extent, the degree to which phenomena are unique is not only greater in geography than in many other sciences, but the unique is of very first practical importance. Hence

generalizations in the form of laws are useless, if not impossible, and any prediction in geography is of insignificant value. For Kant geography is description, for Hartshorne it is 'naive science' or, if we accept this meaning of science, naive description,

(Schaefer, 1953, p. 239)

Schaefer maintained that objects in geography are not more unique than objects in other disciplines and that a science searches for laws. Having eliminated some of the arguments against the concept of a rigorous scientific geography, Schaefer sought to set down the kinds of laws geographers ought to seek. He also urged them to study systematic rather than regional geography.

Hartshorne (1955, p. 242) delivered a very strong counterattack on Schaefer in which he maintained: 'The title and organization of the critique lead the reader to follow the theme of an apparent major issue, "exceptionalism", which proves to be non-existent. Several of the subordinate issues likewise are found to be unreal.' Hartshorne admitted to having used the words **idiographic** and **nomothetic**, but rejected the idea that different sciences can be distinguished as being either idiographic or nomothetic. These two aspects of the scientific approach are present in all branches of knowledge (*ibid.*, p. 231). As early as 1925 Sauer had suggested that, although geographers had earlier been devoted to descriptions of unique places as such, they had also been trying to formulate generalizations and empirical laws.

Both Hettner and Hartshorne made a distinction between **systematic geography** (which seeks to formulate **empirical generalizations** or laws) and the study of the unique in **regional geography** (whereby generalizations are tested so that subsequent theories may be improved). Hartshorne (1959, p. 121) suggests that geographical studies show 'a gradational range along a continuum from those which analyse the most elementary complexes in a real variation over the world, to those which analyse the most complex integrations in areal variation within small areas'. James (1972, p. 468) emphasizes that there is no such thing as a 'real region'. The **region** exists only as an intellectual concept which is useful for a particular purpose. Later critics have read a much more metaphysical significance into the concepts 'unique' and 'region' than was intended by the geographers who were practising between the wars. Incorrect quotations from Hettner and Hartshorne have, however, gained an amazingly wide acceptance: 'It is discouraging to find some writers who continue to accuse Hettner and his followers of defining geography as essentially idiographic, thereby obscuring the underlying continuity of geographic thought' (*ibid.*, p. 228). James thus maintains that what has been called the '**quantitative revolution**' did not represent such a major change in direction as many think.

It can hardly be denied, however, that the interwar generation of geographers were sceptical of the formulation of general and theoretical laws, partly as a reaction against the crudities of environmental determinism. Arguments for idiographic rather than nomothetic approaches seemed to justify the scientific character of studies of the individual case.

ABSOLUTE AND RELATIVE SPACE

Harvey (1969) argued that the concept of geography as a chorological science of the individual case was not tenable because it built on the assumption of **absolute space**. Space in this sense is only an intellectual framework of phenomena, an abstract concept which does not exist in itself independent of objects. Werlen (1993, p. 3) suggests that it is a frame of reference for the material aspects of social actions in the sense of a formal-classificatory concept. In this sense space cannot have any explanatory power (Fosso, 1997, p. 16).

In a practical, classificatory sense, absolute (Euclidian) space is, however, rather useful, but it may be argued that 'faced by the seductive utility of Euclidean space we have allowed an interest in maps to become an obsession' (Forer, 1978, p. 233). Space is in this way treated as a container; first we delimit a spatial section of the earth, say the Newcastle region, and then start to examine its content. The notion of **vertical connections**, humanity's dependence upon local natural resources, was a conceptual basis for such studies. Good examples are found in the French school of regional geography (Box 3.2; Figure 3.7).

Box 3.2   The changing space relations of an English village

Figure 3.7   Rothbury

The rapid changes from vertical to horizontal connections during the second half of the twentieth century are illustrated from Rothbury (Figure 3.7), a small service centre of 2,000 inhabitants, 50 km north-west of Newcastle upon Tyne in an upland farming region. In 1950, Rothbury and its valley were substantially self-sufficient. Most people living in the village found work locally, all the nationally provided services were available in the village and most goods could be bought there. Now Rothbury is governed from Alnwick, 25 km away, and the high school is in Morpeth, also 25 km away. The courts and the railway station have closed and the future of the hospital is not certain. A recent survey showed that 38% of the villagers did little or no shopping in Rothbury and 80% of clothes and hardware shopping was done outside the village. One third of the work-force now drive over 25 km to work and a further 10% work from home, most using computer links. Five local businesses either design software, provide local computer facilities or sell their products nationally via the Internet. Village economy is now largely dependent on high levels of car ownership and on telecommunications.

Figure 3.8 A demonstration of the plasticity of space. The four maps have been constructed from data on the New Zealand airline system and its changes from 1947 to 1970. The two maps on the left show how distance measured in time has changed as the airline network has grown and the speed of travel has increased. The maps on the right show how the net distance travelled has changed with the network
Source: From Forer, 1978

By the end of the nineteenth century, however, the traditional self-sufficient economies of Europe were giving way to an international market economy, and the value of this type of regional study was reduced. **Horizontal connections**, state and international policies, market forces, the interplay between regions, cities and countries, became more important for local development than the local connections between humanity and the land. This was (as noted earlier in Chapter 2) realized by Vidal de la Blache and it led to greater interest in horizontal, spatial structures.

By introducing the concept of **relative space**, horizontal, spatial relations and distance measured in different ways could be given explanatory power. Distance could be measured in terms of transport costs, travel time, mileage through a transport network and even as perceived distance: 'The shift to a relative spatial context is still in progress and is probably the most fundamental change in the history of geography as it opens an almost infinite number of new worlds to explore and map' (Abler *et al.*, 1972, p. 72). One important aspect of this is that geographical features such as settlement patterns, land use, diffusion processes, etc., demonstrate a location and dynamics that are due, to a large extent, to their relative positions in space.

Pip Forer (1978, p. 235) has observed that since distances in time, cost or even network mileage are partly artifacts of socioeconomic demands and technological progress, these types of spaces are naturally dynamic and truly relative. This leads him to the definition of **plastic space** – a space that is continuously changing its size and form. An illustration is given with his own time-space map of New Zealand (Figure 3.8) (*ibid.*, p. 247).

A DISCIPLINE RIPE FOR CHANGE

It was only after the Second World War that theoretical considerations on the relativity of space (and also research issues such as the study of diffusion models and location theory) came to occupy a dominant position in geography. One factor in the adoption of the **'new' geography** was the critical institutional situation many departments of geography found themselves in, particularly in the USA. In 1948, James Conant, president of Harvard University, had reportedly come to the conclusion that 'geography is not a university subject' (Livingstone, 1992, p. 311). The Department of Geography at Harvard was closed soon after, and the discipline was also gradually eased out of some of the other more prestigious private Ivy League universities.

Among the practitioners of the ever more theoretical sciences, the claim that the regional synthesis constituted geography's essential identity lent the subject a dilettante image. After the Second World War the North American universities were expected to produce problem-solvers or social technologists to run ever-increasingly complex economies (Guelke, 1978, p. 45), and geographers were not slow in adopting theory building and modelling that might promote the status of their science and their own academic standing.

Gould (1979, p. 140) recalls how the new generation of geographers were sick and ashamed of 'the bumbling amateurism and antiquarianism that had spent nearly half a century of opportunity in the universities piling up a tip-heap of unstructured factual accounts'. Morrill (1984, p. 64) claimed that the young generation's vision, although it might seem radical to those satisfied with an inferior status for the discipline, was in fact conservative in the sense that 'we wanted to save geography as a field of study and join the mainstream of science'. Even though, or perhaps due to this, the 'new' geography of the 1950s and 1960s was spearheaded by the Americans. They were also inspired by earlier theoretical works in Europe that, so far, had been almost overlooked.

The situation was much less critical in Britain because of the very strong and independent position geography held in both schools and universities. In many of the US states geography was more or less absent from the curriculum as a discrete discipline at high-school level, and less than 1% of students entered universities to study geography. Geography graduates had to find career outlets in applied research and planning. The continual threat of departmental closure or staff reduction based on independent evaluations of research productivity also explains the frenetic search in American universities for new ideas and research programmes.

We now proceed to look at the development of the **spatial science school** and assess whether the changes in geography (also called the **'quantitative revolution'**) really were a scientific revolution in the Kuhnian sense.

THE GROWTH OF SPATIAL SCIENCE

Location theory originates from economic theory. The classic location theories, including Johan Heinrich von Thünen's work on patterns of agricultural land use (1826) and Alfred Weber's study of industrial location (1909), are economic theories. Later economists and regional scientists, including Ohlin, Hoover, Lösch and Isard, developed theories of the areal and regional aspects of economic activity further. **Regional science** developed in some universities as a separate discipline; in yet others, this research came to be linked with economic geography or regional economics.

Walter Christaller (1893–1969) was the first geographer to make a major contribution to location theory with his famous thesis *Die Zentralen Orte in Süddeutschland* (1933), translated by Baskin as *Central Places in Southern Germany* (1966). Christaller, who had studied economics under Weber, declared in 1968 that his work was inspired by economic theory. His supervisor when he was working on *Die Zentralen Orte* was Robert Gradmann, a geographer who had himself made an outstanding regional study of southern Germany (1931) which, however, closely followed the current idiographic tradition in German *Länderkunde*. Although Christaller's thesis was accepted, his work was not appreciated during the 1930s, and when Carl Troll (1947) wrote a review of what had been going on in German geography between the wars, he did not even mention him. In Kuhn's terminology, Christaller's attempt to explain the pattern and hierarchy of central places by a general theoretical model was not acceptable within the reigning paradigm. Christaller never held an official teaching position in geography (Box 3.3; Figures 3.9–3.11).

Eventually Christaller gained a following, notably in North America and Sweden when it was realized that his central place theory could be applied to the planning of new central places and service establishments (Figure 3.11) and also to the delimitation of administrative units. Edward Ullman (1941) was one of the

---

Box 3.3   Christaller's Central place theory

A theme in **landscape morphology** (see Chapter 2) – the morphological network of central places in southern Germany, as seen on the topographic map – was the starting point for Walter Christaller when he, as a 40-year-old PhD student, developed his **central place theory**. He started to 'play with the maps', connecting towns of the same size with straight lines until his map was filled with triangles (Figure 3.9). These triangles appeared to show some regularities. If the region had really been a flat plain with uniform rural population densities, it would seem that the morphological features could be idealized in a hexagonal, hierarchical structure of urban places (Figure 3.10). Christaller used economic theory to explain the rationality of this morphological pattern.

During the Second World War Christaller was asked to use his theoretical abilities in the planning of new German settlements in eastern Europe. But it was only after the Second World War that central place theory had its first real application in the planning of the newly reclaimed Nord Oost Polder in The Netherlands (Figure 3.11).

Figure 3.9 The geometrical hexagonal landscape of towns in southern Germany from Walter Christaller's classic study of central places made in the 1930s

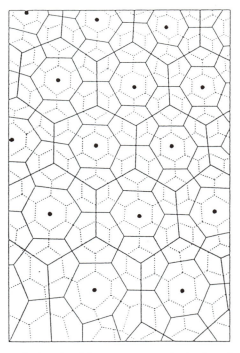

Figure 3.10 When population densities are uneven, the lattice of central places adjusts to the changes, closing up in densely settled areas and opening out in sparsely settled areas.

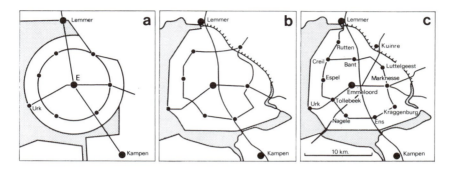

Figure 3.11   Walter Christaller's central place theory applied in the planning of the settlements in Nord Oost Polder in the Netherlands. (A) Geometrical diagram of the proposed settlement pattern; (B) plan of five new villages around Emmeloord; (C) the revised plan as executed
Source: After Meijer, 1981

first American geographers to draw attention to Christaller's work – American geographers were beginning to develop the theoretical models of urban structures and cities as central places that had been devised earlier by economists and urban sociologists (Harris and Ullman, 1945).

In an account of geography as a fundamental research discipline, the influential American geographer Ackerman (1958) encouraged students to concentrate their attention on systematic geography, cultural processes and quantification. A range of different statistical methods was gradually brought into use in several systematic branches of geography, enabling the development of more refined theories and models.

The acceleration of theoretical work was especially marked in institutions led by geographers who had studied the natural sciences, especially physics and statistics, and/or where there were good contacts with developments in theoretical economic literature. During the 1950s at several American universities, the frontier between economics and geography became very productive of new ideas and techniques.

A seminar for PhD students in the use of mathematical statistics conducted by William L. Garrison at the University of Washington, Seattle, from 1955 onwards was of particular significance. Garrison and his co-workers were mainly interested in urban and economic geography, into which they introduced **location theory** based on concepts from economics with associated mathematical methods and statistical procedures (Garrison, 1959–60). Many of the students from Seattle became leaders of the 'new' geography in the USA during the 1960s, including Brian J. L. Berry, William Bunge and Richard Morrill. Both Berry and Garrison later moved to work in the Chicago area. Through the inspiration of Berry the geography department at the University of Chicago became a leading centre of theoretical geography. It attracted a large number of PhD students and published a well-known series of monographs. Berry and other leading professors later left the department (which was closed in 1986), demonstrating the vulnerability of geography on the American academic scene. In the 1950s there was a simultaneous development of theoretical geography at the universities of Iowa (where Schaefer had taught until his death in 1953) and Wisconsin.

It is possible, as Johnston (1997, pp. 62–73) maintains, to recognize in this period four schools of quantitative geography in the USA. Three were developed in the departments of geography at the Universities of Washington, Wisconsin and Iowa, with Washington the most prominent centre of innovation. The fourth – '**social physics**' school – developed independently, drawing its inspiration from physics rather than economics. Its leaders were John Q. Stewart, an astronomer at Princeton University, and William Warntz, a graduate in geography from the University of Pennsylvania (who was later employed as a research associate by the American Geographical Society).

Empirical studies indicated that the movement of persons between two urban centres was proportional to the product of their populations and inversely proportional to the square of the distance between them. Stewart pointed out the **isomorphic** (equal form or structure) relationship between this empirical generalization and Newton's law of gravitation. Thereafter this concept became known as the **gravity model** (Box 3.4). Stewart's ideas about isomorphic relations between social behaviour and the laws of physics were introduced to geographers by a paper in the *Geographical Review* as early as 1947. Here Stewart (*ibid.*, p. 485) stated that human beings 'obey mathematical rules resembling in a general way some of the primitive "laws" of physics'. Warntz, working with Stewart, also borrowed analogy models from physics in his studies of population potentials (Warntz, 1959; 1964). He suggested that the mathematics of population potential is the same as that which describes a gravitational field, a magnetic potential field and an electrostatic potential field (James, 1972, p. 517).

Box 3.4   The gravity model

Early in the nineteenth century, some scientists suggested that the laws of physics could be applied to the study of human relationships and that the laws of gravitation might explain patterns of travel and trade between places. By the mid-twentieth century gravity models were widely applied within the spatial science school of geography. In its simplest form, the **gravity model** can be expressed as follows:

$$I_{ij} = \frac{k\,(P_i \cdot P_j)}{(D_{ij})^2}$$

where I_{ij} represents the interaction between town i and town j; P_i and P_j are the populations of the two towns; D_{ij} *is the distance between them;* and k is a constant.

The equation indicates that the interaction between the two towns (numbers of telephone calls, flows of traffic) is proportionate to the *product* ($\cdot$) of their populations, divided by the *square* – ( )^2 – of the distance between them.

The work of Christaller, August Lösch and others was introduced into Sweden by Edgar Kant, an Estonian geographer who had tested their theories in his homeland before taking refuge in Lund after the Second World War (Kant, 1946; 1951). His research assistant in 1945–6 was Torsten Hägerstrand, a brilliant young geographer who was also working on migration processes. Through his contacts with the Swedish ethnologist Sigfrid Svensson (who had made a number of studies of the relations between innovation and tradition in rural areas using the currently accepted methodology), Hägerstrand became interested in the possibilities of investigating the **process of innovation** with the aid of mathematical and statistical methods. In focusing on the **process**, Hägerstrand made a clear break with the current regional tradition. His dissertation 'Innovations – förloppet ur korologisk synpunkt' (1953, later translated by Pred, 1967, as 'Innovation diffusion as a spatial process') examined the diffusion (or spread) of several innovations among the population of a part of central Sweden. Some of these innovations concerned agricultural practices, such as bovine tuberculosis control and pasture improvement, and others were more general, such as car ownership. With the aid of the so-called 'Monte Carlo' **simulation**, which involves the use of random samples from a known probability distribution, he was able to construct a general **stochastic model** of the process of diffusion. Stochastic literally means at random; stochastic or **probability models** are based on mathematical probability theory and build random variables into their structure. Models may be classified as either stochastic or deterministic. In **deterministic models** the development of some system in time and space can be completely predicted, provided that a set of initial conditions and relationships is known.

The stochastic Hägerstrand model enabled the spread of innovation to be simulated and later tested against empirical study. It was demonstrated that the form of distribution at one stage in the process would influence distribution forms at subsequent stages. Such a model *could* therefore be of use to planners in support of future innovations they wished to bring about. The department of geography at Lund University soon became renowned as a centre of theoretical geography, attracting scholars from many countries. Almost from the beginning there were contacts between Lund and Seattle. Hägerstrand taught in Seattle in 1959 and Morrill studied with him in Lund, where his work on migration and the growth of urban settlement (1965) was presented.

In the years that followed, Hägerstrand's technical and statistical procedures attracted more attention than his theoretical analyses. He himself regarded his work as less important for its empirical findings than for its general analysis of the **diffusion process**. He stated in the first sentence of the dissertation that, although the material used to throw light on the process relates to a single area, this should be regarded as a regrettable necessity rather than a methodological subtlety (Hägerstrand, 1953 – 1967, p. 1). This was of course meant as a deliberate provocation to the traditionally bound regional geographers. Hägerstrand regarded his (1953) analysis of individual fields of information and their change through time as his most important contribution to geographical thought, as the study of such information fields is basic to a deeper understanding of the processes of diffusion.

During the 1960s, Hägerstrand went on to make detailed studies of individual behaviour, using three-dimensional models to portray the movement of individuals in time and space. An important feature of **time-space geography** is that time and space are both regarded as resources that constrain activity. Individuals have different possibilities of movement in space, conditioned by their economic status and technical possessions, but time imposes limitations on everyone. Subsequent studies in time-space geography, which have been carried out actively at Lund and elsewhere throughout recent decades (see, for instance, Carlstein *et al.*, 1978), have shed much new light on geographical aspects of human behaviour.

The 'new' geography spread over the world from the innovative centres, but it did not have the same impact in all countries. Christaller's work had aroused little interest in his home country, Germany. His theories had to take a detour into the English-speaking world – from whence they returned steeped in the 'new' geography – to be fully appreciated. Initial forms of quantification, such as the use of frequency distribution scattergrams, parameters and index numbers, were first applied around 1960. The introduction of **factor analysis**, notably in a classificatory study of Swiss cantons by Steiner (1965), was the real introduction to new quantitative approaches for most German-speaking geographers. The philosophical implications of **the spatial science school** were first presented by Dietrich Bartels in his book *Zur wissenschaftstheoretischen Grundlegung einer Geographie des Menschen* (1968).

Another reason for the delayed impact of the 'new' geography in Germany was that geographers tended to follow Troll's appeal in *Erdkunde* (1947) in that they accepted the principle that worldwide research should be a normal component of an academic career. Virtually all established geographers were attracted to employing their talents abroad, leaving their graduate students to cultivate research at home. Research abroad undoubtedly contributed to Germany's international reputation, particularly through the application of German methods of detailed landscape studies and cartographical work. German geographers also found that their existing techniques were well adapted to research abroad, particularly in the Third World where the statistical basis for quantitative analysis was sparse or unreliable.

WHAT KIND OF REVOLUTION?

Throughout the world there was marked opposition among established geographers to the learning and teaching of the new spatial science methods, and a reluctance to open professional journals to contributions, the editors did not understand. 'There was something electrifying about tilting with the dragons of the establishment' says Morrill (1984, p. 59), and for this reason the young generation of geographers had the feeling of being revolutionaries. In the USA the lack of publication outlets led to the establishment of a theoretically orientated journal, *Geographical Analysis* (*ibid.*, p. 65).

In 1963, a Canadian geographer, Ian Burton, arguing that what he labelled the **'quantitative revolution'** was over and had been for some time, cited the rate at which schools of geography in North America were adding courses in quantitative methods to their requirements for graduate degrees. It must be stated, however, that most geographers did not consider the theoretical developments within the subject as a revolution, and that many 'revolutionaries' were at pains to emphasize continuity in the ultimate objectives of human geography. The use of statistics for the making of relatively precise statements was generally accepted,

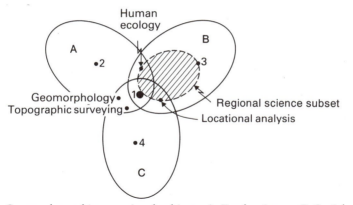

Figure 3.12 Geography and its associated subjects, A: Earth sciences, B: Social sciences, C: Geometrical sciences 1: The core of geography, 2: Geology, 3: Demography and other social sciences, 4: Topology with other geometric sciences
Source: After Haggett, 1965

although the related use of mathematics in modelling received much less attention (Johnston, 1978).

Most research workers regarded advanced statistical methods as being useful in some branches of the discipline; other branches, notably historical and cultural geography, felt less need for new techniques. Leonard Guelke (1977b, p. 3) claimed that 'To an extent that is not widely recognised, the move to quantification took place within the basic framework of geography put forward by Hartshorne in "The nature of geography" (1939)'. In many geography departments the works of both Hartshorne and Garrison were on the students' reading lists, but philosophical and methodological differences between them were not an issue in teaching up to the mid-1960s; Schaefer's criticisms had been forgotten.

Johnston (1997, pp. 74–5) points out that the leaders of the quantitative school did not study the philosophy they were adopting very deeply – apart, that is, from references to the works of Gustav Bergmann, a philosopher and close friend of Schaefer who had actually read the galley proofs for Schaefer's paper on 'exceptionalism' (Schaefer, 1953), contained in some of the papers of the Iowa group and, most notably, in William Bunge's thesis *Theoretical Geography* (1962, 2nd edn, 1966). Bunge, who had worked at Iowa for a short period, extended the arguments of Schaefer to the effect that geography is the science of spatial relations and inter-relations, geometry is the mathematics of space, and so geometry is the language of geography. The **chorological viewpoint**, emphasizing the character of and inter-relationships within specific places or regions, was rejected in favour of a geography based on **spatial analysis**, which stressed the geometric arrangement and the patterns of phenomena. **Relative position in space** – distance measured in various ways – became the main explanatory factor.

With some irony, Bird (1993, p.11) comments that 'the one and only revolution in geography' took place in June 1966: 'An event in the literature enables us to date the last straw, when the last idiographic bastion in geography was overthrown – the destruction of the idea that locations could never be anything but unique.'

In an article Grigg (1965) had tried to argue that all parts of the earth's surface are unique. If it is held that geography includes the study of the location of its data and that 'locations are unique', then geography cannot fully employ the scientific method. In June 1966, however, Bunge published a short commentary to Grigg's paper, asserting that 'locations are not unique', but general. Locations are comparable as witnessed by such terms as 'near', 'far', 'close', 'distant' and 'adjacent', which describe the relativity of locations. It is thus the **relativity of spatial locations** that can be analysed in a scientific way.

Quantification as such does not lead to any scientific revolution in the Kuhnian sense. The change from **absolute** to **relative space** as the focus of geographical study had, however, basic philosophical implications and was in this sense revolutionary. It is therefore better to talk of the **spatial science school** rather than **quantitative revolution** and **quantitative geography** to describe the new trends of the 1950s and 1960s (Figure 3.6 p. 68).

The major advances towards a unifying methodological and philosophical basis for the spatial science school were made in the 1960s by British geographers, notably Peter Haggett, Richard Chorley and David Harvey. *Locational Analysis in Human Geography* by Peter Haggett was published in 1965. The importance of this book lay in its overview of much new theoretical work in the subject. Haggett (*ibid.*, pp. 14–15) used the diagram reproduced in Figure 3.12 to illustrate the argument that there are three traditional subject associations of geography: with the earth sciences (geology and biology), with the social sciences and with the geometrical sciences. Haggett (*ibid.*, pp. 15–16) maintained that:

> The geometrical tradition, the ancient basis of the subject, is now probably the weakest of the three. Much of the most exciting geographical work in the 1960s is emerging from applications of higher order geometries . . . Geometry not only offers a chance of welding aspects of human and physical geography into a new working partnership, but revives the central role of cartography in relation to the two.

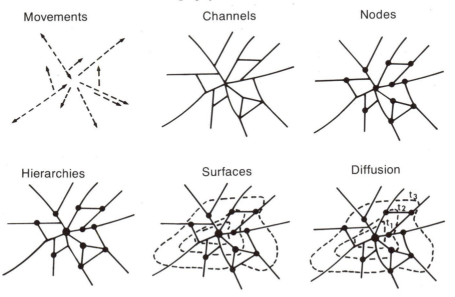

Figure 3.13   The basic elements in Haggett's model for the study of spatial systems, t_1, t_2 and t_3 representing stages in diffusion.
After Haggett, Cliff and Frey 1977 p. 7.

At the heart of geography as a science is the distributional view. Geography is a **discipline in distance**. When we discuss space it is not the container space 'that frames the totality of a landscape; we prefer to think of space as a system of distance relationships between objects' (Hard, 1973, p. 184). The study of spatial arrangements may be summarized in Haggett's (1965) diagram (Figure 3.13) of spatial structures. The sketch may be seen as a disaggregation of **functional regions** such as those established around central places in a Christaller model, into five geometrical elements (movements, channels, nodes, hierarchies and surfaces). A sixth element, diffusion, was added later (Haggett *et al.*, 1977). In contrast to the traditional system of self-sustained regions, the primary element in a modern society is the need and desire for interaction between places which results in a pattern of **movements**. These might be studied as the geometric pattern of straight lines between points, but in fact most movements are channelled along particular route corridors, such as roads. So we can study the patterns of **channels** which, together with **nodes**, represent an organization **network**. The **hierarchy** represents the relative importance of the nodes and the **surfaces** represent the system of land use as exemplified by the work of von Thünen. Patterns of human occupance are, however, not static. The process of change in time therefore involves **spatial diffusion** as developed by Hägerstrand.

Haggett's book led to a fundamental debate within the subject. The arguments presented by Kuhn (1962/1970a) on paradigm shifts within the world of science were applied to the debate. Thus Chorley and Haggett (1967, p. 39) stated that they had looked at the **traditional paradigmatic model** of geography and had found that it was largely classificatory and under severe stress. They suggested that geography should adopt an alternative **model-based paradigm**, and so made it clear that the new development within the subject not only represented a wider range of methods but also demanded a fundamental paradigm shift. Each geographer was given the choice between the traditional and the new model-based paradigms. Model building was set up as the aim of geographical investigation, a task to be performed with the aid of **quantitative methods** and the use of computers to handle data. A **model** was defined as an idealized or simplified representation of reality that seeks to illuminate particular characteristics. The concept is a wide one – for Chorley and Haggett (*ibid.*), a model could be a theory or a law or a hypothesis or a structured idea.

The rapid development of spatial model building and the use of quantitative techniques could not have taken place without computers, but computers did not determine the development of spatial science: 'Model building preceded the invention of the computer in many sciences, but in a discipline like geography which handles such large quantities of data it would hardly have been possible to develop operational models worthy of the name without computers' (Aase, 1970, p. 23). This technological development had given the subject new possibilities young researchers had no hesitation in exploring.

We may conclude that the technical development of the discipline (use of computers and mathematical-statistical methods) could hardly be called a 'revolution' in Kuhn's sense. To talk about a 'quantitative revolution' may thus give false impressions. But it is true that the 'new' geography provided the discipline with notable research projects that could serve as **'exemplars'** for new students as, for instance, Hägerstrand's diffusion model. And the renewed discussion on the basic problems of the subject that followed in the wake of the quantification process may be regarded as a sign of a crisis phase. Individual research workers felt themselves more or less obliged to take a stand and to clarify their own research situation, so there was little opportunity for straightforward **puzzle-solving**. The transformation to a **spatial science** on the basis that **locations are essentially relative** may also indicate a paradigm shift. But the meaning of this transformation was not generally understood or appreciated by the geographical community. Old ideas continued to flourish and new ideas cropped up as results of criticisms of the spatial science school. Bird (1993, p. 13) has characterized the changes in geography as **constant revisions**; they may also be regarded as a multiparadigmatic development, since different schools of thought continued to live side by side.

It may, however, be a characteristic of social science that new paradigms do not become so well established to enable a relatively long period of normal science. Or rather, we may have reached a stage of mature science where we experience **'revolution in permanence'**, in the Popperian sense.

CRITICS OF THE SPATIAL SCIENCE SCHOOL

Opposing the so-called 'revolution', Stamp (1966, p. 18) preferred to call it a 'civil war', and noted that quantification had many points in common with a political ideology; it was more or less a religion to its followers, 'its golden calf is the computer'. Broek (1965, p. 21) stated that 'there are more things between heaven and earth than can safely be entrusted with a computer'. Even Ackerman, one of the advocates of quantification, warned (1963, p. 432) that 'the danger of dead end and nonsense is not removed by "hardware" and symbolic logic'. Minshull (1970, p. 56) observed that the landscape was becoming a nuisance to some geographers, that many of the models could only be applied to a flat, featureless surface, and warned there was a real danger that these ideal generalizations about spatial relationships could be mistaken for statements about reality itself.

Fred Lukermann (1958) reacted especially to attempts by the social physics school to establish analogies with physics, maintaining that hypotheses derived by analogy cannot be tested: falsification is impossible. In a series of papers in the 1970s, Robert Sack, a student of Lukermann, criticized the view put forward by Bunge (1962) and Haggett (1965) that geography is a spatial science and that geometry is the language of geography. Sack (1972) maintained that space, time and matter cannot be separated analytically in a science concerned with providing explanations. The geographical landscape is continuously changing. The processes which have left historical relics and which are creating new inroads all the time must be taken into account as important explanatory factors. The laws of geometry are, however, static – they have no reference to time. The laws of geometry are sufficient to explain and predict geometries, so that if geography aimed only at analysis of points and lines on maps geometry would be sufficient as geographical language. But, 'We do not accept description of changes of its shape as an explanation of the growth of a city . . . Geometry alone, then, cannot answer geographic questions' (*ibid.*, p. 72).

Another problem, Broek pointed out (1965, p. 79), crops up if we project a model derived from our own surroundings over the whole world as a universal truth and measure different situations in other countries as 'deviations' from the 'ideal' construct. Models based on research within the western cultural world cannot be elevated into general truths. Brian Berry (1973b) came to the conclusion that a universal urban geography does not exist, and that urbanization cannot be dealt with as a universal process: 'we are dealing with several fundamentally different processes that have arisen out of differences in culture and time' (*ibid.*, 1973b, p. xii). He divided the world into four universes: (1) North America and Australia, with their free market economies; (2) western Europe, with its planned welfare economy; (3) the Third World, with its economy split between a traditional and modern sector; and (4) the socialist countries, with their rigidly planned economies. Each of these has its own urban geography, which again will change through time.

Haggett *et al.* (1977, p. 24) also noted that 'the Russian translation of the first edition of this book (Haggett, 1965) made clear how heavily the locational explanations were rooted in the classical economics of the capitalist world. Inevitably, the lopsidedness of the book will appeal to certain readers and condemn it to others.'

THE ACHIEVEMENTS OF SPATIAL SCIENCE

It is commonly agreed that the spatial science school threw open the windows of a hitherto introverted discipline, which had had its major links with idiographic disciplines, such as history and geology. Disciplinary boundaries became much more open; methods and theories were openly borrowed from geometry, physics and social sciences as geographers became involved in multidisciplinary research projects. The 1960s and 1970s were optimistic decades for geographical innovators. Student numbers grew rapidly and career opportunities expanded considerably.

The redevelopment of geography as a social science raised the self-esteem of geographers and opened up a job market for candidates within planning and administration. Generalists (as geographers still were, but now with added technical and statistical knowledge) proved to be better adapted to the job market than candidates with narrower specializations.

Haggett (1990, p. 6) argues for practical and pragmatic approaches in geography: if 'science is the art of the soluble, then much geography is the art of the mappable'. 'Thinking geographically', liking maps and thinking by means of them is intrinsically linked with geography. More than any other natural or social science, geography is a **visual science** with similarities in this respect to architecture and the history of art. We like to climb a mountain in order to get an overview, a grand survey of the geographical patterns in front of us. We try to describe and explain the world as we perceive it.

Description and mapping were also central to the traditional schools of geography, but the spatial science school developed more refined methods that made spatial correlations and statistical tests possible. The most recognizable shift was, however, the downgrading of ordered description of what we know (**cognitive description**) and the development of sophisticated models and methods in **morphometric analysis** such as are described by Figure 3.13 (p. 84). Most of the models created were simplifications of spatial morphological patterns based on empirical data. Christaller's central place theory is an example of this. When trying to achieve a general, theoretical explanation of patterns, theory was imported from other sciences. In many cases it was economic theory.

A major achievement of the spatial science school has been the development of sophisticated methods for the detection of spatial patterns. Many of the models, including such a simple one as the 'gravity model', are good devices to compare data and thus to describe geographical differences. These approaches have given valuable insights into the geographical patterns which form the bases of our analysis or are the results of our decisions. But it might be argued that spatial science research developed greater refinement of description than explanation. Many commentators within human geography have pointed out that spatial science research has been confined to the empirical level, and that we need **structuration theory** to understand how 'real' or deep structures influence the empirical outcomes or events. (We will return to this in Chapters 4 and 5.)

But still, in the frenetic search for grand explanations, we often forget the value of descriptions that enlighten us. New descriptive models are certainly legitimate scientific endeavours as long as they create new knowledge. Spatial analysis provided better tools for descriptions, and new, intriguing developments have continued to be developed, particularly within physical geography and ecogeography.

Haines-Young (1989, p. 31) points out that 'the new information-based technologies' provide techniques, notably expert systems, which enable us to carry the problem of modelling geographical knowledge to a deeper level than has been possible so far. For example, a system designed to predict fire risk in the Kakadu National Park in Australia uses a database together with information supplied by the users of the park (see GIS, pp. 180–182).

Advanced **systems analysis** has proved its usefulness in **physical geography** and **ecogeography** – the study of humanity's role in changing the face of the earth. Gregory (1985), Goudie (1990) and Huggett (1993) provide many examples of these developments, and earlier editions of this book (Holt-Jensen, 1981; 1988) also included a basic introduction to systems analysis in geography.

Here we restrict the presentation to one example (see Box 3.5 and Figure 3.14 on the Sahel catastrophe), and agree with Unwin (1992, p. 129) that it is 'surprising that ecosystems were not more extensively used as a framework for empirical research by geographers'. One reason for this deplorable fact may be that biogeography has generally held a weak position in Anglo-American and Scandinavian geography; however, the situation is much better in Germany.

The potentialities of **systems analysis** in general, and **ecosystems analysis** in particular, have not been fully developed. Further, systems analysis could also be applied to human geography, as Mabogunje (1976) has demonstrated: it is well suited to giving a conceptual understanding of the factors that influence for instance, rural–urban migration in developing countries.

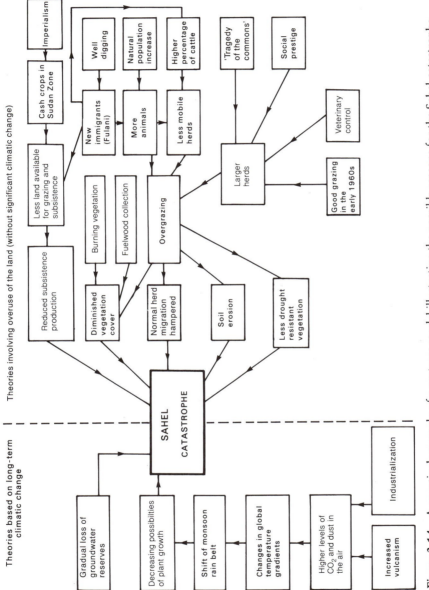

Figure 3.14 A practical example of a systems model illustrating the possible reasons for the Sahel catastrophe
Source: After Reenberg, 1982

Box 3.5   The Sahel catastrophe explained with the aid of a systems model

Chorley (1973a, p. 167) suggests that it is important that the challenge of environmental deterioration is met with new positive feedback in the form of better planning and control – a task for geographers. Let us consider the Sahel catastrophe, which is analysed in a *systems model* (Figure 3.14) by the Danish geographer Reenberg (1982). She uses the model to illustrate the great number of possible causes for the increasing desertification of the Sahel belt of Africa. One possible explanation is climatic change (on the left side of the figure). Climatic changes may be due to strictly natural factors that have prevailed many times in the earth's history. These changes may even have had specific causes, such as increased vulcanism. Alternatively, climatic change may also be induced by human activity, through increased industrialization and pollution. The assumption that increased desertification is the result of climatic change can be tested empirically by studying a series of meteorological observations taken over time.

Having studied the Sahel problem, however, geographers prefer the theories presented on the right side of the model. It is not possible to postulate a simple chain of causes and effects on this side of the model since many of the elements intermingle and work on each other in complex ways. Some of these elements are associated with the region's traditional ways of life. For example, many tribes calculate their wealth in terms of head of cattle, and so tend to overgraze. Other traditions include slash-and-burn methods of improving pasture and agricultural land, the collection of fuelwood and the production of charcoal.

While external influences (labelled 'imperialism' in the figure) were relatively small, some sort of natural balance was struck, which was regulated by mortality in periods of drought. However, external influences have encouraged the population to use the more fertile parts of the region for cash-crop production for sale in more industrialized countries. Consequently, areas of subsistence agriculture have been reduced and the pressures on these areas increased. Well meaning western initiatives, such as medical provision and veterinary control, have meant that the population of both people and cattle has increased, further exacerbating the problem of pressure on the land. Development aid in the form of well-drilling to provide supplies of clean water has also influenced the nomadic population's wandering routes, resulting in patchy overgrazing around watering places. Vegetation is destroyed and the desert expands day by day. A particular problem is the 'tragedy of the commons'. Although not continuously occupied in former times most land was held in some sort of tribal ownership. Western administrators, however, often classified land that was not currently being used as 'common land', effectively removing local tribal responsibility for this land which resulted in a loss of ecological balance.

From this somewhat simplified example we can appreciate the complexity of the Sahel problem, which does not lend itself to mere cause-and-effect explanations or to general structural theories. In undertaking a research project we may not be able to analyse more than a few factors; hence the systems model remains conceptual but, as such, can help us to understand complex relationships.

A 'CRITICAL' REVOLUTION?

In 1972 Haggett appeared confident that the spatial science school had taken the lead: 'Today the general acceptance of [quantitative] techniques, the more complete mathematical training of a new generation and the widespread availability of standard computers on campus make the conflicts of a decade ago seem unreal' (*ibid.*, p. 460). He also pointed out (*ibid.*) that 'the first years of over-enthusiastic pressing of quantitative methods on a reluctant profession have given way to the present phase in which mathematical methods are just one of many tools for approaching geographic problems'.

In the meantime, a new type of criticism had been developing. In *Directions in Geography* (Chorley, 1973b) a number of geographers who had, one way or another, had some hand in the spatial science innovations discussed possible directions the discipline might follow in the future. Many of the contributors suggested quite new directions for research, while others openly criticized different aspects of the spatial science approach. Harvey (1973, pp. 128–9) also became a notable apostate, declaring that 'the quantitative revolution has run its course, and diminishing marginal returns are apparently setting in . . . Our paradigm is not coping well . . . It is ripe for overthrow'.

In retrospect, the 1960s could be characterized as an era of **'hard science'** whereas, in the 1970s, there was much questioning of the law-seeking approach. Guelke (1977a, p. 385) concluded that 'the idea that a scientific discipline must necessarily have laws of its own is false. A discipline can be scientific if it uses or consumes laws from other areas.' Michael Chisholm (1975, pp. 123–5) noted that geographical 'laws' in general would not meet the exacting specifications needed to qualify as laws, since they are not generally verifiable. It would be more correct to talk of models and theories rather than laws in geography, as a **theory** should be defined as an articulated system of ideas or statements held as an explanation (*ibid.*, pp. 123–6).

In Chisholm's view the essential characteristic of central place theory, and other theories established by the spatial science school, is their **normative** character; the theoretical construct is not intended to show how the world is actually organized but to demonstrate the patterns that would occur if reality were rational.

Descriptive, or what Chisholm called **positive, theories** seek to account for observed phenomena, as did, for example, Copernicus' theory of the motion of the planets around the sun. The urban rank–size relationship is one of the more famous regularities observed in geography, and may as such qualify as a positive theory, according to Chisholm (*ibid.*, p. 148). In positive theories the observations of discrepancies between the predicted state and actual states of the system may stimulate an advance or changes in the theory, whereas **normative theory** is used to create a world that is 'rational'. Normative theories are useful in social studies and in town and regional planning, and it was in these fields that many trained geographers found career outlets in the 1960s and 1970s. It was the many assignations in this field, observes Olof Wärneryd (1977, p. 29), that encouraged the situation where geography lost contact with its earlier scientific traditions. As in other fields of social activity, there was a general belief in economic growth and economic theories; however, the values embedded in these were not seriously debated.

Another problem was that many of the geographical models used in planning were static – for example, **central place theory** played a major role in many development programmes but little attention was given to the fact that functional space is dynamic and in more or less continual change.

An example from Sweden, discussed by Gunnar Olsson (1974), may clarify this point. In the 1960s Swedish geographers were engaged in a far-reaching reform of administrative districts, which was expressly intended to abolish spatial social and economic inequalities in the country. The new municipalities were to be large enough to sustain the considerable burden of the municipal welfare services. The methodology used was to observe how the majority of people interacted in space and then to translate the observations into a Christaller-type model:

> Unnoticed by spectators and performers, the play was changed in the middle of the act. The ought of justice disappeared into the wings, invisibly stabbed by the is of the methodology. Exit man with his precise visions, hopes and fears. Enter Thiessen polygons with crude distance minimisations and cost-benefit ratios.
>
> *(Ibid.,* p. 355)

No one thought to ask whether people wanted to change their observed interaction patterns or whether the centralization led to disadvantages for some groups.

The spatial science of the 1960s made use of theories from other sciences, notably economics, which were thought to give an objective description of society and how it functions. Models were constructed which gave an apparent explanation, but they were misleading or directly fallacious in so far as they failed to recognize that social situations might be amenable to change: the models that were developed tended to support society's existing conditions. This is particularly true of process models, which encourage us to believe that a trend, once ascertained, will continue to operate in the future. Thus a theory which appears to be positive or based on realistic description is translated into the development's purpose through planning and political decision. It is during this sequence of events that the theory becomes normative. In the Swedish example, a planning decision that had the objective of changing a social structure in fact led to the conservation of the structure because of the theories and methods used in the planning process. One conclusion that may be drawn from this is that the distinction between normative and positive theory may be useful in a theoretical classification but we must not ignore the fact that the same theory can be used in both a normative and positive way.

Another conclusion drawn by a number of geographers in the late 1960s was that physical planning had not been as effective in fostering social change and equality as many people had hoped. For example, many of the land-use and transport plans that spread from North America to practically every large city in western and northern Europe, and in which many trained geographers had participated, seemed to have increased the segregation of social classes and to have sharpened differences in mobility between the car-owning and the car-less groups. In transport planning, the interaction pattern of the average family had been used as a guideline. Such 'deviant' travel patterns as those of old people with no access to a car had not been given much attention. The reason for this had been methodological; quantitative models were built to cope with aggregate and **'hard' data** – that is, data easily expressed in numbers. **'Soft' data**, which concern

'TRADITIONAL'	'QUANTITATIVE'	'CRITICAL'

Historical geography ————————————————————————➤

Cultural geography ————————————————————————➤

                                Humanistic geography ———➤

Landscape studies ————————————————————➤

Regionalization ————— Computer mapping – GIS ——————➤

Population and settlement ————————————————➤
studies

                Central place theory ————————————➤

Economic geography ————————————————————➤

                Location theory————————————————➤

                                Behavioural geography ———➤

                Social physics————————————————➤

                Innovation studies————————————➤

                                Time–space geography———➤

                Systems analysis————————————➤

                                Ecogeography ————➤

                                Welfare geography——————➤

                                Radical geography——————➤

         In current practice —————➤

Figure 3.15 Schools of human geography 1950–1980. It must be stressed that this classification should be treated with caution, for the criteria for the classification are not consistent throughout the scheme. While 'traditional' schools are mainly classified on the basis of their themes, most, but not all, of the 'quantitative' schools are classified in terms of their type of theory. The 'critical' schools are a real mixture: some are classified on the basis of their special methodology, while others owe their position to a particular political approach. Systems analysis is better regarded as a general method which might be useful to a number of schools rather than as a school in itself. The expressions 'traditional', 'quantitative' and 'critical' are in inverted commas because the use of these terms as labels is highly debatable – as is the grouping of schools under each label

human attitudes and deviations in behaviour, could not be handled easily in such models. But even research workers involved in aggregate studies were bound to wonder about deviations from the 'normal', and this led to studies of the welfare of special groups of people, such as old people, and a growing concern for the position of the individual within a mass society.

Students of **location theory** also had second thoughts as they came to realize that **'economic man'**, that decision-maker blessed with perfect predictive ability and knowledge of all cost factors, does not in fact exist. Locational decisions may be made on a rational basis, but this rationality relates 'to the environment as it is perceived by the decision-maker, which may be quite different from either "objective reality" or the world as seen by the researcher' (Johnston, 1997, p. 150). It was thus necessary to develop alternative theories to those based on 'economic man' and to investigate the decision-makers' behaviour and perceptions. There are a number of different research trends within geography that stem from these observations. Figure 3.15 gives a simplified classification for the period between the Second World War and the 1980s. The 'traditional' and 'quantitative' schools have been dealt with earlier. Some of the trends within 'critical' geography will be outlined briefly below; a discussion of the philosophical basis of 'critical' geography is postponed to the next chapter.

A NEW HUMANISTIC GEOGRAPHY

The new concern for the individual led some geographers to favour a revitalization of some of the methods of the traditional schools in geography, including those of the French school of regional geography. This trend, which may be regarded as a new initiative by geographers who had not been involved in the models-orientated approach of the 1950s and 1960s, emphasizes the need to study unique events rather than the spuriously general. Anne Buttimer (1978, p. 73) argued along similar lines to Vidal de la Blache that historical and geographical studies belong together. She stressed the need to understand each region and its inhabitants from the 'inside' – that is, on the basis of the local perspective and not from the perspective of the researching 'outsider'. Leonard Guelke (1974, p. 193) advocated an idealist approach which 'is a method by which one can rethink the thoughts of those whose actions he seeks to explain'.

Guelke mainly based his arguments on the writings of R. G. Collingwood, the Oxford historian and philosopher, and especially on *The Idea of History* (1946). Guelke argued that these ideas are crucial to **historical geography**; it is important that historical geographers 'focus their attention on the meaning of human actions of geographical interest, not merely their geographical [physical] expressions' (Guelke, 1982, p. 12). 'Different people in making use of the Earth have, for example, created distinctive field and settlement patterns. These patterns are not arbitrary, but reflect the thinking of the people who created them' (Guelke, 1981, p. 132).

We must not be constrained by the view that geographers are always obliged to use the methods of natural science. A natural scientist is normally an outside observer: it is difficult for him or her to take an inside view. Although a historian, in Collingwood's sense, is unable to see directly into the minds of his or her subjects, he or she is able to understand their actions as being products (as are the historian's) of **rational thought**. The historian will therefore use the **hermeneutic** interpretative approach (see Chapter 4), involving *verstehen* or sympathetic understanding – trying to rethink the thoughts of historical characters, as far as his or her knowledge of their cultural background allows, but not trespassing into psychology. When we understand the beliefs which encouraged Columbus to sail west to reach India, we have explained the motives for the voyage. It is not necessary to discuss why Columbus held those beliefs for that would take us into the realm of historical psychology.

The **idealist approach**, as advocated by Guelke, thus restricts itself to beliefs that are in some way rational, and it leaves aside the emotional and psychological aspects of human behaviour. We cannot re-experience the emotional life of other people. The approach, however, does not let us know for sure whether we have really succeeded in finding the true explanation for historical events. The idealist philosophy has elements in common with **phenomenology**, which has also been suggested as a useful approach for geographers.

Phenomenologists do not, however, make sharp distinctions between intellectual and emotional life. The phenomenologist is more concerned with describing the life experience of the researcher. This description must inevitably be subjective because 'the approach fails to distinguish those elements of human existence that are open to subjective understanding and those that are not' (Guelke, 1981, p. 144). Phenomenologists are primarily concerned with our (subjective) **knowledge** of the world around us: they attempt to create an objectified knowledge of our experience of the phenomenal world. Our knowledge proceeds from the world of experience and cannot be independent of that world.

Yi-Fu Tuan, although not defining himself a phenomenologist in the strict sense, has written a number of inspiring essays and books introducing geographers to phenomenological ideas (1974; 1976; 1977; 1980). Tuan has stated (1971) that geography is *the mirror of humanity*: to know the world is to know oneself. Such a study is clearly based in the humanities rather than in social and physical sciences: 'The model for the regional geographers of humanist leaning is . . . the Victorian novelist who strives to achieve a synthesis of the subjective and the objective' (Tuan, 1978, p. 204). Tuan prefers to use the term **humanistic geography** for such studies, which are regarded here as including both idealism (in Guelke's terminology) and phenomenology.

In his book *Conceptions of Space in Social Thought* (1980) Robert David Sack rejected those conceptions of space that stem from the natural sciences and that were the only ones used in the spatial sciences. He suggests that, in other modes of thought such as art, myth, magic and the child's view of life, space may have very different meanings: 'If people see and/or evaluate things and space differently and "non-scientifically", then social science must somehow represent and capture those meanings' (*ibid.*, p. 8).

Such humanistic trends of thought may have been new to the English-speaking world, but they can be related to significant traditions in both French and German geography. Ewald Banse, whom Hettner described as the *enfant terrible* of geography, attacked the ruling disciplinary matrix of the subject as early as the 1920s, arguing that it only attempted to describe external reality. To understand the essence of things in depth, says Banse (1924, p. 58), the science of geography should be redefined as an art: 'Only the unified perspective of both the visible outer appearance and the inner core of things constitute real geography, which is a spiritual presentation of experienced impressions.'

Nicholas Entrikin (1976, p. 616) makes the point that the humanist approach is best understood as a form of criticism. However, as Johnston (1997, p. 196) points out, we may also hold the view that 'the human condition can only be indicated by humanistic endeavour, for attitudes, impressions and subjective relations to places (the "**sense of place**") cannot be revealed by positivist research'. This can provide insight into the essential structures of human relations to the environment.

It is also clear that we will always use and need **hermeneutic** or interpretative methods in geography. Dorling and Fairbairn (1997) point out that even a map is an interpretation of reality and has to be reinterpreted according to the purposes of each particular use.

SPATIAL SCIENCE CRITICIZED FROM WITHIN

In contrast to humanistic geography, **behavioural geography** may be seen as being based on criticism from within spatial science. This starts from a disillusion with **location theories** involving, for instance, the concept of '**economic man**'. However, the roots of behavioural geography are much older. In Europe, the Finnish geographer Johannes Gabriel Granö and his Estonian student Edgar Kant were attempting a behaviourist approach in the 1920s (Granö, 1929). Even in the USA, behavioural and quantitative approaches were contemporary developments: the behavioural approach was taken up in the late 1950s and the 1960s by Gilbert White, then at the University of Chicago, and his associates, who made a series of investigations into the human response to natural hazards, guided by theories of decision-making and influenced by methods used in psychology and sociology. It was regarded as more important to map the personal **perception** of the decision-maker than to describe the factual physical and economic conditions of the environment, since the decision-maker would act upon his or her own perceptions and not on the environmental factors themselves (White, 1973). Julian Wolpert introduced behavioural geography to many human geographers through a paper in 1964, which compared actual with potential labour productivity on farms in central Sweden. He found that the sample farm population did not achieve profit maximization and nor were its goals solely directed to that objective. The farmers were 'spatial satisfiers' rather than 'economic men'.

A further aspect of behavioural analysis is the concept of the **mental map** of the environment. Mental mapping has been taken up by a number of workers, among them Rodney White and Peter Gould (1974). A somewhat different approach to behavioural work is found in Allan Pred's (1967; 1969) two-volume work *Behaviour and Location*, in which Pred tried to present an ambitious alternative to theory-building based on 'economic man'. The **time-space geography** Hägerstrand and his associates (see p. 81) established may also be seen as a critique not so much of the spatial science school as of important aspects of social science research in general. In simple terms, it provides a method of mapping spatial behaviour while at the same time representing a reorientation of scale away from aggregate data towards studies of individual behaviour. More important, however, is its introduction of a new economic theory in which time and space are regarded as scarce resources, the allocation of which forms the basis of the social realities we study.

A corresponding concern for the individual within mass society is also basic to **welfare geography**, which developed as a special branch in the 1970s. Paul Knox (1975) stated that it was a fundamental objective for geography to map social and spatial variations in the quality of life. The study of such spatial inequalities was taken up by Coates and Rawstron (1971), Morrill and Wohlenberg (1971), Coates *et al.* (1977) and Smith (1979), and a number of others. While some of this work represented the geographer as a 'delver and dove-tailer', a provider of information, other examples, notably *The Geography of Poverty in the United States* (Morrill and Wohlenberg, 1971), also proposed both social and spatial policies for changing existing conditions.

Whereas, in principle, welfare geography works within the framework of the existing economic and social system, **radical geography** called for both revolutionary theory and revolutionary practice. Its aim was clearly voiced by Harvey in *Social Justice and the City* (1973, p. 137): 'Our objective is to eliminate ghettos. Therefore, the only valid policy with respect to this objective is to eliminate the conditions which give rise to the truth of the theory. In other words, we wish the von Thünen theory of the urban land market to become not true.' Harvey maintained this could only be done if the market economy was eliminated. The task, therefore, was the self-conscious construction of a new paradigm for social geographic thought, which might stimulate a political awakening and start a social movement with the ultimate goal of bringing about a social revolution.

Berry (1974) commented that Harvey relied too much on economic explanations; in our postindustrial society, control is no longer economic but political, making a Marxist analysis more or less passé. Morrill (1974), however, in reviewing Harvey's *Social Justice and the City* (1973), confessed that he was pulled most of the way by the revolutionary analysis, but that he could not make the final leap in accepting that our task is no longer to find truth but to create and accept a particular truth.

REVOLUTION OR EVOLUTION?

Just as Harvey favoured revolution in society, so he supported the **paradigm** concept and the simplified Kuhnian model of the development of science through revolutions: 'A quick survey of the history of thought in social science shows that revolutions do indeed occur' (Harvey, 1973, p. 122). Harvey assumed that the **motivation** for the construction of a paradigm in social science is the desire to manipulate and control human activity and social phenomena. The **spatial science paradigm**, according to this view, came into existence as a response to pressures from the material, or economic, base of society, which suggested the desirability of discovering ways of manipulation and control, particularly within the planning sector (Quaini, 1982, p. 155).

Harvey (1973, p. 121) did not regard Kuhn as a revolutionary because of 'his abstraction of scientific knowledge from its materialistic base'. According to Jan Widberg (1978, p. 9), Kuhn provides a dialectic-idealistic interpretation of scientific advancement. Widberg maintained that it is possible to describe the history of scientific thought within a discipline on the basis of four fundamentally different views. These are categorized through two sets of dichotomies:

	Mechanical	Dialectical
Idealistic	I	II
Materialistic	III	IV

A **mechanical view** implies that the development of science is linear, with each new generation continuing from the point where the old generation left off. There are no revolutions, only growing specialization, professionalization and the advancement of better methods and understanding. Kuhn expresses the **dialectical view** when he maintains that a science develops through contradictions and revolutions with changes of paradigms.

An **idealistic viewpoint** implies that ideas are the driving force behind scientific development and change. Each individual makes his or her own choice, and it is the genius of scientists that counts. A **materialistic viewpoint** implies that the material base governs the advancement of scientific knowledge. Scientific activity reflects the special interests of those who are in control of the means of production (*ibid.*, pp. 2–3). In Harvey and Widberg's telling, the story of geography has been the tale of the geographical legitimation of the social conditions that produced it.

But it is also possible to argue that in geography paradigms (or, rather, schools of thought) have not succeeded each other but, to a greater extent, continue to exist in parallel. New schools slowly absorb the older ones, leaving some former contradictions to linger on within the new structure. Figures 3.6 and 3.15 suggest that concepts and lines of thought survive after basic shifts have taken place in the discipline, and may crop up again in new clothing.

Anne Buttimer (1981, p. 82) maintains that the idea of a 'paradigm' has proved appropriate for describing developments within the physical sciences; it fits less comfortably in the story of biological sciences and finds itself on rough ground when applied to any field that aims at a comprehensive understanding of humanity and environment. Buttimer (*ibid.*) maintains that everyone who has studied in some depth the life, work and experiences of those individuals who have taken a leading role in the shaping of our science will appreciate that the paradigm model gives a distorted picture of actual developments.

Figure 3.15 illustrates the existence of parallel schools of thought within geography since the Second World War. This is in line with Johnston (1997, p. 389), who concludes that 'human geography is currently characterized by a multi-paradigm situation at the world-view level – and by a wealth of exemplars on which research is based'.

From this we may draw the conclusion that the dichotomy between a dialectical and mechanical understanding of the history of our subject is an interesting academic study, but the truth as we see it lies somewhere between the contradictions. Shifts of major importance do occur, but they seldom encompass the whole scientific community – old ideas and concepts remain with us to a large extent; new discoveries may sometimes have the character of mutations – and usually they look more like rephrasings of old truths.

To analyse the dichotomy between an idealistic and a materialistic view, we have to discuss social research and values more closely. Initially this means we need to get a better understanding of the debate on **positivism** and **critical philosophy**.

4

Positivism and its Critics

The prestige and support that aid research stem from the general belief that research is an almost objective, value-free activity. Most people think of research as an uncompromising search for truth, an activity unaffected by the emotions, beliefs, attitudes and desires of either the researcher or of the society in which the research takes place. When research workers discuss scientific results in the same way as politicians discuss political issues, many people will conclude that there must be something wrong with the research and hence its prestige sinks. Social scientists who work with less exact or less clearly definable data are particular targets for such attitudes.

Natural scientists, as a rule, work with discrete materials which lend themselves to experiment and objective analysis. Their conclusions are widely received as truths. Social scientists have a weaker database, are often personally involved in the research issue and can seldom experiment with their material. The question of objectivity is therefore most acute for social scientists. The results of social science research are often regarded by the public as if they were merely points of view. A discussion on the role of values in the social sciences is therefore important – not only for the viability of research activity but also for relationships between scientists and the general population, and for public attitudes towards science.

POSITIVISM AND CRITICAL THEORY

Scientific and philosophical discussion has produced two chief categories of meta-theory (superior theories or theories about theories) for scientific research: **positivism** and **critical theory**. 'Both trends are in a sense more "climates of opinion" than definite schools of thought. There is far more discussion within the trends than between them', says Skjervheim (1974, p. 213). There are a number of different versions and definitions for each of these meta-theories.

Critical theory is associated with a group of scholars frequently known as the Frankfurt School, represented (particularly since 1950) by Jürgen Habermas. As we do not restrict the discussion below simply to the work of the **Frankfurt School**, we will henceforth use the expression 'critics of positivism' rather than 'critical theory'. Positivism is usually identified with the school of **logical positivism** as it was elaborated by the so-called Vienna Circle in the 1920s. Logical positivism was intended to be a renewal of two closely related traditions in the philosophy of science: the British **empiricism** of John Locke (1632–1704) and David Hume (1711–76) and the continental **positivism** of Auguste Comte (1798–1857). There are, however, other philosophers and scientists who belong to the empiricist and positivist traditions who do not accept the particular developments of the ideas adopted by the Vienna Circle of **logical positivists**. There have, of course, also been further developments in positivism since the 1920s, and so positivism here will be considered in its wider context. Karl Popper (1902–94), although associated with the **Vienna Circle**, was regarded by this circle as 'our official opponent' because he criticized their use of the **verification** principle as a central focus of positivist methodology, developing the alternative **falsification** principle. As we have discussed earlier, Popper has been considered as the founder of a new methodology of science called **critical rationalism**, which can be regarded as an alternative to logical positivism. Following Johnston (1986a) we will include *critical rationalism* in our discussion of positivism. Leaving the thorny semantics of scientific philosophy, we will now try to clarify the development and leading principles of positivism.

THE DEVELOPMENT OF POSITIVISM

After the Renaissance when a range of taboos against **empirical** research had been dispelled, science (notably in seventeenth century England) developed as an independent school of thought, separate from religion. Locke formulated the empiricist principle that all knowledge is derived from the evidence of the senses; and what is not derived from the evidence of the senses is not knowledge. Reliable knowledge can only come from basic observations of actual conditions. To be scientific is to be objective, truthful and neutral.

Comte, who later defined **positivism** as a scientific ideal in line with Locke's principles, believed that alongside the natural sciences there should also be a science of social relationships (which he called sociology) that should be developed on the same principles as the natural sciences. As natural sciences uncovered the laws of nature, so scientific investigation of communities would uncover the laws of society. He admitted that social phenomena are more complex than natural phenomena but believed strongly that the laws governing society would eventually be discovered and hence subjective elements in research would be eradicated. This belief is central to Comte's proposition that social development takes place in three stages: 1) **theological** when people explain everything as God's will; 2) **metaphysical** or free-based speculation (as characteristic of the 'Enlightenment' movement – see below); and 3) **positive** when causal connections are revealed between empirically observed phenomena.

Comte coined the term **'positivism'** in 1830s' France to counter what he regarded as **'negative'** aspects of the pre-Revolutionary **'Enlightenment' movement**. The Enlightenment, associated with Voltaire, Rousseau and other French thinkers, stressed the importance of universal education and the value of individual common sense. Through enlightenment the people were to be liberated from superstition and barbaric rule. This philosophical tradition also involved a romantic and **speculative** approach that sought to change society by considering **Utopian** alternatives to existing situations. Comte regarded this Utopian speculation as 'negative' since it was neither constructive nor practical; it demonstrated that philosophy was an 'immature' science. Philosophers, like other scientists, should not concern themselves with such speculative matter, but should study things they could get to grips with: material objects and given circumstances. This approach was to be recommended as '**the positive approach**'.

Comte himself wanted to change society's development but stated that the nature of positivism is not to destroy but to organize. He argued that organized, gradual development and change should replace the disorder created by the French Revolution. Comte identified free speculation, or **systematic doubt** – as defined by Rene Descartes (1596–1650) – as the **metaphysical principle**. **Metaphysics** was defined as that which lies outside our sense perceptions or is independent of them. Positivists should regard metaphysical questions as unscientific. Comte held the **metaphysical principle (systematic doubt)** responsible for the French Revolution, which, having begun to tear down the feudal structure of society in a state of emotional enthusiasm, had ended in despotism. In a positive society scientific knowledge would replace both religious belief and free speculation. In this way positivism was progressive in that it strongly opposed the belief that the Bible was authoritative in scientific inquiry.

As was shown in Chapter 2, the publication of Darwin's *Origin of Species* was a major boost for positivism as a scientific ideal. The great stress laid by positivism on empirical data and replicable research methods enabled a marked development of science during the nineteenth century. Because metaphysical questions came to be regarded as unscientific, science developed its own objectives that were apparently free of belief and value postulates. Positivism tends to be **anti-authoritarian** in so far as it requires us not to believe in anything until there is empirical evidence for it, and it can be investigated by controlled methods. We should not therefore accept authority because it is authority, but only give credence to things for which there is scientific evidence. This sceptical attitude, and the consequential search for certainty, has naturally enough brought conflict for positivists and led them into confrontation with dictatorial regimes.

This was particularly the case with the **Vienna Circle of logical positivists**, founded in the 1920s. Its members were opposed to everything that smacked of metaphysics and unverifiable phenomena. They therefore became bitter opponents of Nazism, which they saw as a mixture of irrational prejudice and ideological dogma. 'Positivist' became a term of abuse in Nazi Germany and was applied to Alfred Hettner, among others (van Valkenburg, 1952, p. 110). The Nazis wanted research to be based on their own ideology. Moritz Schlick, the leader of the Vienna Circle, was murdered, and most of the Circle's other members were driven abroad. As long as science is uncompromising in its search for the truth it will threaten regimes that are based on systematic lies and ideological postulates.

In defining the rigorous principles that science must follow in order for it to be called science, the Vienna Circle added the adjective '**logical**' to positivism. The methods of **formal logic** had to be used to define the elementary statements or

axioms that would constitute fundamental empirical knowledge and to derive further statements of knowledge from those axioms. **Euclidian geometry** (geometry as you learnt it in school) provides a classical example of this process. A few fundamental axioms provide the basis for a system of geometrical knowledge proved through a formal scientific verification procedure. New, verifiable statements or hypotheses can be formulated from fundamental propositions and from formerly proven statements or laws. These new hypotheses must be tested in turn against reality to establish their veracity (Johannessen, 1985, p. 59).

PRINCIPLES IN POSITIVISM

A major aspect of logical positivism is its emphasis on the **unity of science**. Scientific status is guaranteed by a common experience of reality. A common scientific language and method ensure that observations can be repeated. Since science has a unified method, there can only be one comprehensive science. The common method is the **hypothetic-deductive method** and the model discipline is physics. The language that will make a unification of science possible is the physical language or **'thing' language**. The ultimate aim is, in the words of Rudolf Carnap, to construct 'all of science, including psychology, on the basis of physics, so that all theoretical terms are definable by those of physics and all laws derivable from those of physics' (cited in Skjervheim, 1974, p. 222). The poles and the system of latitude and longitude are the only special definitions that must be made before pursuing geographical research. It follows from this that disciplines are to be distinguished from each other by their object of study, and not by their method (Gregory, 1978, p. 27).

From empiricism, positivism derived the central thesis that science can only concern itself with **empirical questions** (those with a factual content), and not with **normative questions** (questions about values and intentions). Empirical questions are questions about how things are in reality. In this context 'reality' is defined as the world that can be sensed. This means that science is concerned with objects in the world. The subject, or subjects for which there is a world (or worlds), are excluded from the field of interest.

Disregarding the question as to whether what we can sense as objects comprises the whole of reality, it could be said that empirical questions are about what a thing is and normative questions are about what a thing should be: 'How are the available food resources distributed between the inhabitants of the world?' is an empirical question. The corresponding normative question would be: 'How should the available food resources be distributed between the inhabitants of the world?'

Positivism holds that, since we cannot investigate such things as moral norms with our senses, science should keep away from normative questions; we cannot justify our tastes scientifically. Science can describe how things are and, experimentally or by some other measurement, discover the association of causes that explain why things are as they are. The research worker can, given his or her knowledge of contemporary associations of causes, forecast possible future developments from given propositions. But science cannot from 'is' statements draw conclusions about 'should' statements. Ideally, science is value-free, neutral, impartial and objective. When the scientist evaluates things, using 'should' statements, he or she is no longer a scientist but becomes a moralist or a political campaigner. In this way it can be said that positivism also displays the characteristics of an ideology, in so far as it claims that its particular mode of inquiry is the sole valid approach to scientific knowledge and that other modes are metaphysical and non-scientific (Johnston, 1986a, p. 18).

CRITICISMS OF POSITIVISM

Because positivism seeks authority from the methods of natural science, it is not so anti-authoritarian as it claims to be. This may lead positivists into thinking that there are technical solutions to all problems – an essentially conservative standpoint. The wish to be free from value judgements may lead scientists to build ivory towers for themselves, wherein research takes place without much discussion of its objectives and implications. Positivist research workers might restrict themselves to describing how things are and how they will develop if they continue on the same track as now. Critics maintain that the prestige modern science has acquired creates an aura of inevitability around these 'are' statements and tendencies.

Others go further in arguing that value-free research is impossible. Subjective elements will intrude in many stages of the research process, especially at the stage when research workers choose a topic for study from the many available. We can, for example, imagine that research workers, starting from their own well established and strong opinions as to what the distribution of the world's food supply should be, will choose to investigate the empirical question as to *how* the food supply is distributed. Even if the researchers do not deliberately consider *what* the distribution should be, it would be difficult for them to exclude their own views wholly at the stages of problem formation and the interpretation of results. It is also obvious that once results are available, the description of the existing distribution will influence the views of decision-makers as to what the distribution should be. In this process we can say that scientific activity is itself shaping reality. It is no longer a passive observer.

Other critics maintain that the positivists' belief in the unity of science is wholly unrealistic; positivism allows its view of the logic of science to influence its conception of the content of science. We should not oversimplify things by laying down rules as to how science should function without taking account of what actually happens within the livelier research traditions. The fact is that several main lines of scientific inquiry (including important schools in psychology and social science) do not show any sign of developing towards the ideal of a unified science: ideas about unified systems of concepts belong on the drawing boards of scientific theoreticians; the inner drift in science itself is to give each area of investigation its own form of expression.

It is even possible to argue that each and every scientific school of thought is a form of cognition, an agreed approach to the analysis of the world. Teaching a discipline consists in teaching its current forms of cognition. When, for example, one learns to see things geographically, it is not reality itself one learns but a perspective on reality. What the positivists are trying to do is to base all knowledge on methods that have been carried over from the natural sciences into the human sciences. They have developed a specifically technical conception of science, and they try to exclude scientific traditions that do not adapt to their recommended technical language and methods. A typical example is Schaefer's rebuttal of the chorological viewpoint and his efforts to make geography into a **spatial science** (see p. 72–73).

Wilhelm Dilthey (1833–1911) considered that while we *explain* nature, we *understand* social life and human intentions. This appreciation lies at the root of the **hermeneutic** (Greek *hermeneuin* = to 'interpret') tradition within social science that, by putting oneself into another person's shoes, tries to reveal, for instance, human intentionality by *verstehen* (a German word which means something between understanding and empathy).

We should note here that the English-speaking peoples use the term 'science' in the context of a positivist approach following the model of the natural sciences rather than those of arts and humanities. The German expression *Wissenschaft* is much wider and comprises all forms of methodological study carried out in a systematic way and based on defined theoretical assumptions about the nature of research objects (Johannessen, 1985, p. 153). There are at least two main types of *Wissenschaft* that are separate from and not inferior to each other. Wilhelm Windelband (1848–1915) agreed with Dilthey in rejecting the positivist concept of a unified science, preferring to distinguish between the **nomothetic** (law-seeking) sciences and the **idiographic** (descriptive) or historical sciences. As we have seen, the distinction has had a considerable impact on the debate on geographical methodology (see, for instance, Figure 3.6, p. 68).

The Diltheyan tradition supports the use of empirical methods within the natural sciences but does not accept the insistence of Comte and Carnap that the social/cultural sciences should copy the methods of natural science. When social scientists borrow models of system building from natural sciences they must treat all their elements as objects. When the mind is treated in this way it is materialized either directly (by being conceived of as the thing that thinks) or indirectly (by being considered as a relation between objects in the world). The same applies to behaviour.

At the extreme, we might talk about brain processes instead of ideas or images, but we cannot study human behaviour in the same way as we study animal behaviour. The difference is that humans have intentions. Intentional expressions such as 'to imagine something', 'to believe something', 'to love somebody' cannot be translated into the 'thing' language of the natural sciences: they cannot be understood as objects as seen from the outside. If you study your fellow human beings as physical objects you will not get to grips with their intentions. It is better to build bridges to other people by breaking down the barriers between the observer and the observed and creating an intersubjective understanding. This is the principle of subjectivity in social science, which says that behaviour has to be studied and described in terms of the acting individual's orientation towards the situation.

Immanuel Kant criticized empiricism as a philosophy because, as a man of deep religious belief, he was concerned about what he saw as empiricism's 'nihilistic implications'. Neither empiricism nor positivism leave room for a God, nothing is a priori certain. Kant's need to ordain certain main features of truth was satisfied by his doctrine of 'categories', in which he claimed that if there is no prime cause for the content of reality, the form of reality must be primarily a given. What the well of consciousness is filled with is an empirical question but, regardless of what is put into it, the content of the consciousness is shaped by the form of the well. Human beings cannot know how things are in themselves; we only recognize things as they take shape through our senses and by our conceptualizing them. The reason why pure chaos does not reign within the field of scientific activity is that human perception and reflection are built upon common categories within which impulses are classified (space, time, cause) and, as we saw above, Kant classified the branches of science into three corresponding categories.

DIALECTICS, HEGEL AND MARX

Georg Wilhelm Friedrich Hegel (1770–1831) made scientific inquiry even more difficult by arguing that the categories we use for classification and thought (structure of concepts, speech) are not fixed for all time but are historically and socially conditioned. Often we cannot gain a full understanding of a thing by merely studying it on its own. We must also consider its antithesis – its opposite. In fact, continual consideration of the antithesis may well throw new light on the current thesis. The development of knowledge in the Hegelian model is described as **dialectic**. Increased insight into the new thesis, gained through contrasting it with its antithesis, may lead to a synthesis. This in its turn enables us to set up new theses and antitheses. This process does not achieve a permanent form of knowledge, correct for all time. Knowledge and science are not like a steadily growing ant-hill of pine needles but are defined better as continually changing and deepening processes. This method has in fact been used in Chapter 3 and in this chapter as an educational principle. In Chapter 3 we asked whether geography had developed as a science through gradual evolution or through paradigm shifts and revolutions. The conclusion came as neither-nor, but . . . This chapter sets up positivism against its supposed antithesis and discusses to what extent geographers have followed one or the other.

Gunnar Olsson (1975, p. 29) suggests that the crucial point about dialectics is that understanding and creation themselves are moved by dialectical transitions; truths are relative; and statements can be designated as 'true' only at a given point in time and, in any case, they can be contradicted by other 'true' statements:

> While conventional reasoning knows only the either-or distinctions of the excluded middle, reality knows the both-this-and-that relation of dialectics and many-valued logics. Since contradiction is not external to reality but built into its structure, the language in which reality is discussed should itself have the same characteristics of internal negation. It is in this sense that reality is dialectical, for both reality and dialectics are governed by perpetual processes of internal tensions and not by stultifying juxtapositions of opposites. As a consequence, both the empirical and the logical bases of our theories should be open to conceptual change. It is rather to falsify for the sake of creative understanding. This is possible, because dialectic movement is not inference but deepening of concepts.

(Ibid., pp. 28–9)

Hegel (1975, pp. 153–4) discussed dialectics with special reference to geography. He differed from Kant but agreed with his contemporary, Ritter, in defining geography as a historical rather than as a naturalistic discipline, 'a study of the possible modes of living offered by the environment to people settled in various regions of the Earth' (Quaini, 1982, p. 19). As the historical development of civilization gradually frees people from the constraints of the natural conditions under which they live, a dialectic relationship develops between the geographical environment and its population's way of life. It has been maintained that Karl Marx (1818 – 83), who developed the dialectical method further, was also influenced by Ritter – we know he attended Ritter's lectures – as he turned his thoughts from Hegelian **idealism** to **historical materialism** (Cornu, 1955). In so far as Marx's most important methodological texts are primarily devoted to a critique of Hegel, idealism and the mysteries of speculative construction, they naturally include rather important elements of **positivism**. Marx believed it was necessary to turn from the abstract to the concrete (Quaini, 1982, p. 29). The consciousness was for him more a product of material conditions than their foundation. The material world and humanity's behaviour therein were the **base** upon which thoughts and ideologies formed a **superstructure**.

An aspect common to Marxism and positivism is the **theory of realistic knowledge** as this was developed in early or 'naive' **realism**. Knowledge means cognizance of an objective truth: it regards the world as concrete. **Marxism** also proclaims the unity of science, but the basis for this unity is not so clear cut as in positivism. Engels leaned to the natural sciences when he maintained that the gradually increasing similarity between theories of social science and those of natural science would lead to the accommodation of nature and society within a unified philosophical perspective. In his youth, Marx also wrote about a unified science which comprised nature, society and human psychology. However, while positivists maintained that a unified science should be based upon the methods of the natural sciences, Marx considered the philosophy of the social sciences to be potentially far superior to that of the natural sciences. Therefore the eventual fusion of the two fields of study would come about through the socialization of the natural sciences (Harvey, 1973, p. 128).

Marx asserted that we know only one science, the science of history (Quaini, 1982, p. 35). It also follows from the grand Marxist theory that society develops in stages in accordance with developments in the factors of production. Marx refused to accept that the scientific laws governing society were eternal. This view contrasts sharply with the claim of positivist science that scientific laws are universal in space and time. Engels pointed out that 'to us, so-called economic laws are not eternal laws of nature but historical laws which appear and disappear' (cited in Gregory, 1978, p. 73). Marx was particularly critical of the economists Adam Smith and David Ricardo; he considered their outlook on reality to be unhistoric and non-dialectic. Society has inbuilt conflicts which will resolve themselves by change, both in practice and in theory.

SUBJECTIVITY AND OBJECTIVITY

Critics of positivism maintain that metaphysical assumptions cannot be excluded from science. Facts are not facts in themselves: they represent those parts of reality that can be appreciated with the concept apparatus available. Facts are facts only in relation to a given scientific aim which is itself structured by the values intrinsic in society. The scientific process is therefore restricted by the environment in which it takes place and is constrained within the limits of the research workers' perception. Foucault (see p. 62), for instance, argues that scientific truth is a relative concept, dependent on power relations within societies. Within particular societies and during periods of time power relations create particular **epistemes** (conditions for acquiring knowledge). In his insistence on the relativity of truth, Foucault rejects any form of absolutism or any perspective that will seek to provide a definitive view of society. This leads particularly to a rejection of Marxism (Poster, 1984, p. 39). Both Foucault and Habermas criticize Marx for reducing social practice to labour, and for thus failing to comprehend fully the interconnectivity between theory and practice.

Although some, with Foucault, regard the relativity of truth as valid throughout science, most critics of positivism would stress – like Habermas (1978) (see Box 4.1) – the different practices within natural and social sciences. Natural scientists have not needed to question the empirical-analytic form of scientific investigation. They have been able to construct a free-standing scientific language – the **'thing' language** – which satisfies their need for precision. Admittedly, this language was established through scientists' subjective appreciation of nature, but this creates few problems as long as scientists broadly agree on the experience of the senses. Nature can only answer in the language in which the question is put – which is the language of science. Therefore mistranslations can only occur through the faulty use of accepted methods and language.

Many scientists would argue that much of the criticism of positivism has little relevance to research procedures in natural sciences, having more relevance to the application of science. It may be possible to pursue more or less **objective** research within the natural sciences, but the results are applied by society in such a way that science becomes a tremendous force that transforms the community (Ackerman, 1963, p. 430). Value judgements are implicit in the application of scientific results – in nuclear physics, for example – and consequently natural scientists cannot avoid making up their minds about them. Even the priorities given to different types of research are connected with values; it is not irrelevant whether the priority we give to research funding is in nuclear physics or ecology.

In social science, **subjectivity** (or the problem of values) is deeply involved in both theory and practice. Concepts in social science are related to human evaluation. A typical example is the concept of natural resources. A natural scientist might study coal as a thing in nature, but in social relationships coal is a resource which must be translated as 'something of value to humanity'. Coal is interesting to social scientists because it is a resource and forms the basis for coal-mining districts. But coal has not always been a resource, and not all coals found in nature are resources. The market situation and the cost of extraction may lead to closure of a coal-mine even if there is a lot of coal remaining. The human evaluation that creeps in makes the use of the 'thing' language difficult in social science.

Another problem is that ordinary language must be used widely in the social sciences because it is spoken by the objectives of the research and also because there is a close affinity between daily speech and technical language. When, for example, we ask a commuter for his or her opinions about his or her own travel to work, we must take account of the concepts as understood by the interviewee. A problem arises when we define our concepts 'scientifically' before embarking on a

---

Box 4.1   Different forms of science, according to Habermas

Table 4.1   Habermas's theory of cognitive interests

Form of science	Knowledge-constitutive interest	Social medium	Expression through
Empirical-analytic	Technical	Work	Material production
Historical-hermeneutic	Practical	Language	Communication
Critical	Emancipatory	Power	Relations of domination and constraint

Source: Unwin (1992) based on Habermas (1978)

Habermas maintains that all knowledge is historically and socially rooted and interest bound. This is the basis of what he terms cognitive interests, or knowledge-constitutive interests. There are three types of such cognitive interests (Table 4.1):

1) Knowledge which enables the human species to *control* objects in nature and thus produce the necessities of its material existence. This is the basis for empirical-analytic sciences which leads to technical knowledge and which is connected to the social medium of work. Empirical-analytic (or positivist) science is fundamentally geared to the production of technically useful knowledge, and is concerned with the prediction and control of processes that have been objectified.

2) Knowledge which enables people to *communicate*, to understand each other. This is the basis for historical-hermeneutic sciences, which aim to provide practical knowledge through language that is essential for understanding and communication. Hermeneutics involves understanding on the part of thinking, living individuals, whereas positivism is concerned with explaining a reality that is seen as existing independently of the individual's interpretation. Habermas argues, however, that hermeneutics could not provide a substantial grounding for human action as there is no bridge between the worlds of individual meanings and the worlds of social action. So in fact it does not provide practical knowledge; it only communicates individual human understanding.

3) Knowledge which enables people to *act rationally*, to be self-determining and self-reflective. This is the basis for critical science, which aims at emancipation through the social medium of power. Self-reflection provides the central methodological core of Habermas' conception of a critical science.

These three forms of science lead on to newer trends in social-science methodology (see Box 4.2 p. 112 – on hermeneutics, critical theory and poststructuralism).

Source: Based on Unwin, 1992, pp. 30–1

research project and later use these definitions to interpret our results without evaluating critically what happened during the investigation. The person who has been interviewed may have a wholly different understanding of the concepts. In the social sciences, a dialectic relation takes place between subject and object. It is impossible for the research worker who is investigating social phenomena to regard these as objects wholly external to him or herself. The subject is him or herself part of the object: the social scientist acts as a part of the society he or she studies.

The dialectic between the subject and the object leads on to what has been termed **double hermeneutics**. In their writings, social scientists explain whatever they have understood about the beliefs and attitudes of their fellow citizens, but the people they write for are more or less identical to the people they are writing about. If readers understand what the writer has understood, they may be led to change their attitudes or actions. These new attitudes can then be investigated again by the social scientist, explained and understood *ad infinitum*. The meaning ascribed to the one constantly mediates the meaning ascribed to the other through double hermeneutics (Skjervheim, 1974, pp. 298–9; Gregory, 1978, p. 61).

Since our knowledge of the world is established through scientific concepts, reality and knowledge can be changed through human reflection and practice. Reality, as constituted by scientists, can therefore be transformed into something else. **Critical theory**, as developed by the **Frankfurt school** and by Habermas, suggested that science should take an interest in these possibilities of change. Habermas asserted that the relationship of the social sciences to the community is similar to that of the psychoanalyst to his or her patient. The psychoanalyst treats his or her patient by using his or her own intimate *verstehen* (understanding) of the patient. He or she tries to make the patient understand the underlying causes of his or her problems and encourages him or her to believe that these causes can be altered, converting what the patient believes to be permanent constraints into **pseudo-constraints**. The patient who becomes conscious of these constraining conditions may be restored to health if he or she really wants to be healthy. Social scientists, who claim to be professionally established reflections of society, must also reveal pseudo-constraints so that potential opportunities for change in society are clarified.

For this reason, the Frankfurt school developed '*Ideologiekritik*', which aims to enlighten agents about their true interests. A distinction is made between ideology in the *pejorative* sense and ideology in the *positive* sense. Ideology in the pejorative sense is a system of pseudo-constraints or 'systems of beliefs and attitudes accepted by the agents for reasons or motives which those agents could not acknowledge' (Guess, 1981, p. 20). In general such an ideology makes it possible for a minority group to impose surplus repression on their society – that is, more repression than is needed to maintain that society. The other members of society accept the minority ideology, unaware that the underlying motive of the minority is 'surplus repression'. Positive ideology, on the other hand, represents the desiderata for a particular society, which reflect the true interests of its members. This is something to be constructed, created or invented through discussion and reflection within the social sciences: 'A critical theory, then, is a reflective theory which gives agents a kind of knowledge inherently productive of enlightenment and emancipation' (*ibid.*, p. 2).

The parallel between psychoanalysis and social science could be developed further if we accept a priori that society is sick. Social scientists may then make a conscious value judgement that it is their obligation to cure a sick society. Although there are certain rules for *Ideologiekritik*, it is difficult to define an ideology objectively in the positive sense in so far as the 'agents' do not understand their 'true interests'. This is the point made by Lenin when he argued that 'the correct proletarian world-view must be introduced into the proletariat from the outside by members of the vanguard party' (Guess, 1981, p. 23).

Positivists, on the other hand, argue that adherents of critical theory confuse scientific problems with others of a predominantly human interest. It is possible to describe ideologies and their *raison d'être* scientifically; it may well be possible to unmask the real motives behind an ideology in the pejorative sense. But it is not possible to construct positive ideologies scientifically: this is a task for political resolution. Science should only study things that can be treated objectively; if human intentions are not susceptible to such scientific inquiry, scientists should refrain from analysing them.

Most positivists will agree that value judgements play a major role in the research process and that science ultimately takes part in the shaping of society. Positivism is, however, not a description of how research is carried out; it represents an *ideal* of how it should be done.

PRACTICAL CONSEQUENCES FOR RESEARCH

The chief effect of the positivism debate on practical social research has been to enliven discussion about the influence of value judgements on research activity. The thorough-going positivist standpoint stresses the importance of reducing the value element as far as is possible. Research workers would be wise to refrain from problems on which they hold strong opinions that might influence the research process or cause faults that would reduce the value of the findings. Ultimately, we should avoid totally research themes where strongly divergent views are held so as to escape being implicated in political conflicts.

Others consider we should not avoid contentious questions or even questions in which we have a strong interest; in these circumstances we should be on the alert for outbreaks of subjectivity. This can be achieved by maintaining careful accuracy in description and by carefully describing our methods so that other research workers can confirm our results. Some scientists would not regard this as sufficient. As well as taking care to avoid subjectivity, research workers should express their personal viewpoints clearly so that their significance can be evaluated and accounted for in the overall consideration of the research results.

A contrary view is that research workers should take up exactly those questions in which they feel deeply involved. The influence of value judgements can be positive when it motivates us to greater efforts. Research workers should declare rather than try to eliminate the influence of values on their work, so that these influences, being observed, will enable different views within the research environment to correct each other.

Some would go as far as to say that research workers should use science to fight for their values. If the 'value' is a political viewpoint of the research worker, he or she may set the problem and publish the results of research in order to support his or her political viewpoint. People who think this way argue that research, according to the positivist ideal, does not consider how things could have been, only what they are. Such research supports the existence of, and consequently provides subliminal evidence for, the predominant values in society. The research worker who does not share these values must therefore be free to pursue research with the intention of changing basic values, and therefore the nature of society.

There will undoubtedly be different views as to how we should tackle the practical research problems created by value elements. We may agree that neither research ethics nor objectivism demand freedom from values: the influence of values must be accepted as an inescapable element in research. Research ethics therefore become closely associated with the degree to which value judgements are clarified. Gunnar Myrdal (1953, p. 242) claimed that 'We need viewpoints and they presume valuations. A "disinterested" social science is from this viewpoint pure nonsense. It never existed and it never will exist.'

The practical consequence of such an appreciation is that the research worker must be free to choose research projects he or she regards as critical. This implies that scientists themselves, by and large, should have the power to decide how the funds for research should be distributed. Certainly each community will have a research policy but this should not be carried so far that politicians and those assigning tasks at the administrative level outside the scientific environment actually determine research tasks in detail. This approach to the freedom of research is widely shared among scientists, although in many countries the reality does not conform to this ideal.

Politicians and people in general are, however, just as qualified to make value judgements as are scientists. If science is admitted as a force that transforms society, and the ivory tower of science becomes a governing structure of society, clearly it must be brought under democratic control. It is impossible for science to maintain both privileged autonomy and a commanding position. Until now scientists have generally regarded their autonomy as more important than their power, and have consequently tried to separate their political from their scientific activities. As both a scientist and a politician, the author appreciates that he has been in a privileged position in his political activity, since scientists are generally better trained to formulate problems and to sift out the essential elements from voluminous official documents than is the ordinary citizen. For this reason the influence of science and scientists may be much more of a problem for democracy than the influence of society is for science.

In **empirical-analytic** research, value elements are most significant in the first phases of the research process when research problems are chosen, set and defined; and also in the concluding phases of research when the results are interpreted and presented. In the intervening data-handling phase, value elements do not play such a large role – they concern primarily the classification of data. When interpreting and presenting research the problem arises as to how far to carry the conclusions and where to cut short. Is the assignment completed when the relationships shown by the data have been described, or should we also use our imagination and theoretical insight to add something about the circumstances under which these relationships will change or disappear? This problem is particularly acute for social scientists, where the possibility of changing direction depends on human attitudes and intentions.

Box 4.2   Interpretation and reflection

Whereas **quantitative methods** are well suited to analyse physical systems in which data are well defined, they have to be supplemented or replaced by other methods in the scientific analysis of human activity systems that are poorly structured and in which data are difficult to quantify. Alvesson and Sköldberg (1994, pp. 10–11) point out that a polarized debate on the merits of quantitative versus **qualitative methods** is not fruitful. This question cannot be solved in an abstract discussion, but has to be related to the actual research problem and its objectives. Some cases, even in social science, demand a purely quantitative approach; others require a purely qualitative approach; and often a combination of the two is more satisfactory. The main point is that the scientific problem has to be defined first. Then we might choose the most appropriate methods of solving it (Bordieu and Wacquant, 1992).

Qualitative methods are concerned with interpretation and reflection and on this basis Alvesson and Sköldberg (1994) suggest four types of reflective research:

1) *Empirical approach* (particularly as exemplified by **grounded theory**), which gives weight to systematics and research techniques. Qualitative methods ought to follow a closely considered logical approach in relation to the empirical material and should also use well defined techniques for data handling. The result is an empirical descriptive study in which empirical induction plays a major role, but in which hermeneutic interpretation, critical theory and poststructuralism are also of some importance.

2) *Hermeneutic approach*, in which research is primarily seen as an interpretative activity. It is important to realize that all research includes and is done by an interpreter, who acts together with and studies other interpreters (see double hermeneutics, p. 109). Methods cannot be separated from theory, as preconceived concepts and suppositions guide the interpretations and the results of the study. Here hermeneutics is a predominant form of research thinking, but in a research project other approaches will also play a role.

3) *Critical theory* stresses that the scientist needs to be aware of the political-ideological character of research. Social science is a social phenomenon which cannot be separated from its political and ethical context. Research and the modes of research will have difficulty in avoiding either the acceptance or challenge of existing social conditions. Interpretations and the theoretical assumptions upon which these are based are not neutral but are embedded in and support the construction of political and ideological preconceptions. A research project based on critical theory cannot, however, completely avoid the other approaches.

4) *Poststructuralism* suggests that texts may be seen as having an independent life: a scientific author's text is, in great measure, a reflection of the discourses that provide the context for the text. There is nothing chaotic or anarchistic about the way discourses emerge in particular settings (e.g. institutional), but this mode of thought does, of course, raise questions about the ability of individual geniuses to create wholly new texts (how could we understand such a text?). One important conclusion is that **self-reflection** is crucial for researchers – what role am I playing? How am I perceived by the subjects of my research? An important aspect is the importance of reflection on the power of language and concepts, since

these are interpreted differently by different people. We need, therefore, some understanding of **semiotics**, which deals with symbols, signs and meanings as objects of
research.

When it comes to practical research, the different types of research that have been
described will employ and mix different approaches to interpretation and reflection. In
each type of research, however, some elements are more dominant than others.

Source: Based on Alvesson and Sköldberg, 1994, pp. 14–15

GEOGRAPHY AND EMPIRICISM

The fact that, in its intermediate position within the sciences, geography includes
elements of natural science is one reason for the relative late start of the debate on
positivism. Another reason is the **idiographic tradition** in geography, which has
been supported by geography's strong links with other idiographic disciplines,
such as geology, biology and history.

Geology is the natural science that has traditionally been closest to geography.
The research methods of geology were described *inter alia* by Simpson (1963, p.
46) as different in nature from those of physics. He associated geology with what
he called historical sciences. Geology refrains from the formulation, verification or
rejection of hypotheses through experiments and the establishment of universal
scientific laws. Its approach is **empiricist**, but does not follow the strict scientific
rules of logical positivism. It describes and clarifies 'concrete' simple phenomena
and puts them into a geological chronology and classification.

Similar comments have been made concerning traditional views on history as a
discipline. Dray (1966, p. 45), for instance, argued that 'History . . . seeks to
describe and explain what actually happened in all its concrete detail . . . Since . . .
historical events are unique, it is not possible for the historian to explain his
subject matter by means of covering laws'. In physics individual phenomena and
their combinations cannot be presented as really new objects, but historical
phenomena, either singly or in combination, can hardly ever be subjected to a
uniform form of measurement; they exist as truly new phenomena. The laws of
the natural sciences may be regarded as having unrestricted universal validity but
historical generalizations are not valid for all times and places.

The majority **idiographic** view (based on the significance of individual occurrences) which has dominated these neighbouring sciences has, as we have seen
above, also been important in geography. The **determinism** of the nineteenth
century proved to be rather an unfortunate attempt to establish laws, but its
failure to do so led to *scepticism towards nomothetic approaches*. Geographers therefore turned their backs on that major element of positivism, i.e. the unity of
science with one methodology whose results are not modified by time and space.

Since Darwin's time, however, other elements of positivism have dominated
traditional geography. Geography was clearly defined as an **empirical science**. Its
data were regarded as concrete – things experienced by the senses exist as facts.
The question as to what reality could or should consist of was regarded as not

scientific, being instead political or metaphysical in nature. This view may well present fewer problems for geography than it does for sociology and other social sciences. That part of the database of geography which concerns natural conditions is concrete, and may be defined by natural science terminology and methods of measurement.

Many data in human geography are similar to data in physical geography in that objects which appear on a map or aerial photograph (including patterns of settlement, lines of communication and elements of land use) are as concrete as are the data used by agronomists, geologists and biologists.

Although other material in human geography is not so clearly manifest in the **cultural landscape**, part of it is concrete and measurable. We may consider, for example, the numbers settled or employed at a particular place; transport as measured in quantities of goods carried, or numbers of vehicles or persons; the use of raw materials and energy in production and the amount of material available; and so on. These are data that are in principle as easy to handle as those the 'concrete' natural scientists study but, in practice, human geographers must depend on data collected by others, such as maps and statistical material. This implies that the data are already interpreted before our investigation starts. In an investigation, interpretations crop up at many levels and can create much confusion and insecurity in the minds of research students. As pointed out by Biilmann (1981), all research includes both empirical and hermeneutic (or interpretative) methods (see Box 4.2).

A concrete representation of 'reality', such as a map, is the result of a number of transformations that involve interpretation and human evaluation. The cartographer has to interpret data and transform it for the purpose of the map. The map's scale and symbols may well be decided by someone else, depending on the purpose the map is intended to serve. Axelsen and Jones (1987) even question whether all maps are not, in fact, **mental maps**. The map users, of course, will make their own interpretations depending on their ability to read maps and their level of critical appreciation.

We must often be content with data we have collected through interviews, which may be affected by the perceptions of both interviewers and interviewees. We may use data drawn from statistical publications which may have been collated from a number of sources. In both cases subjective elements can introduce flaws. The definitions of science may be different from those of daily speech. Another problem may arise when officials who prepare the statistical material for publication do not understand or do not use definitions in a proper manner. A third problem is that the definitions change from one census to another (for example, the concept 'household' had quite a different meaning in 1970 than in 1900).

Geography is similar to social anthropology and sociology in its concentration on contemporary data. Geographers, however, usually work with more concrete or quantitative data, whilst sociologists and social anthropologists, to a greater extent, use 'interpreted' data – data about people's values. Sociologists are often concerned with the interviewee's subjective understanding of a value-loaded question. They must make certain the interviewee understands the question exactly so there are no misconceptions. Geographers have this problem too, but many geographical questions are easier to define precisely. Sociologists must decide whether the interviewee's answer expresses his or her subjective meaning objectively. This problem has also been recognized by geographers who have worked with 'soft' interview data. We often interview people who have not thought previously about our question, and who either give the answer they believe will be most acceptable to us or the answer that is in accordance with what they believe to be the general understanding of the question, even if this does not coincide with their own point of view. A young lad who was asked why he had moved from mid-Wales to Birmingham answered the question the easy way by saying there were no jobs in mid-Wales (which was an acceptable answer). He concealed his personal reasons for moving: 'Things are more exciting in Birmingham and there are more girls around.' Gradually geography has become more involved in this type of data which, in the traditional, regional school was regarded as sociological data geographers did not need to address.

The traditional, regional school of geography may be labelled **empiricist** and its methodology could, as pointed out by Hansen (1994, p. 46), be described as **descriptive induction**, or presentation of experienced facts. Bird (1993, p. 45), in a tabular simplification, characterizes the empiricist approach as follows:

- Things experienced by the senses exist as facts. These represent its *ontology*, its theory of existence.
- Knowledge can be achieved by sense experience. This is its *epistemology*, its theory of knowledge.

THE POSITIVISM OF SPATIAL SCIENCE

Quantitative methods may be employed by both critical and positivist schools of thought. However, when geography began to draw on economics to articulate more formal location theories, its own somewhat fuzzy empiricism, observes Derek Gregory (1978, p. 40), was considerably strengthened and sharpened. Spatial science involved accepting elements of positivism that had previously been disregarded – namely, the concept that there is one science and one methodology which extends from the natural into the human sciences. By rejecting the notion that geographical phenomena are singular, the spatial science school discarded the discipline's idiographic traditions and set out to discover universals: to build models and establish theoretical structures into which geographical reality might be fitted.

The more distinguished proponents of spatial science, however, made the essential reservation that their models and laws cannot be understood as if they were laws as in natural science. They are measuring rods – tools used to test departures from geographical reality. Critics like Gregory (*ibid.*), however, argued that even if it had not been possible to show that geographical phenomena are subject to universal laws, there was still some value in regarding them as if they were. Some of the spatial science models and laws, he maintained, had been used as devices whose utility is measured by the success of their predictions and not by their implicit validity or truth. This approach to laws, which may be termed **instrumentalism**, was borrowed by geographers from neoclassical economics in which (according to Gregory, *ibid.*, p. 41), it has played an important supporting role. Instrumentalism refers to the extent to which models and laws are seen as instruments of manipulation rather than explanatory devices. Chisholm (1975, p. 125) referred to the same idea as **normative theory** (see p. 91).

Johnston (1986a, p. 33) maintains that Harvey's book *Explanation in Geography* (1969) – which gave the most thorough representation of the philosophy and methodology of the spatial science school – also presented a basically positivist methodology with strong overtones of **logical positivism**. Johnston (*ibid.*) further associates the more important theories within the quantitative school (including central place theory, land-use theories derived from von Thünen, industrial location theory and spatial interaction theory) with positivism. He agrees (*ibid.*, p. 52) that 'empiricism, positivism and empirical research are not the same thing, and that quantification is not exclusive to any single philosophy. But much research using quantitative methods in human geography has been closer to the positivist than any other approach.'

Livingstone (1992, p. 324) maintains that 'the early history of quantification reveals that social and political interests evidently underlay the enterprise. This of course should not be taken to mean that quantification was *only* ideological crystallization.' Harvey (1984) explained the advance of quantification in American geography as a strategic move to escape the political suspicion falling on social science in the McCarthy era. A technocratic spatial science, crystallized in the application of control engineering to geographical systems, may even be said to have bolstered western capitalism and demoted value judgements.

Bennett (1985), however, makes the point that the critique of geography as positivist relates almost exclusively to human geography, and is generally irrelevant to the important body of geographical work concerned with physical-human relations. To a large extent, the critics who seek to demolish the spatial science school by labelling it positivist also accept uncritically that geography is exclusively a social science. Many of the geographers who worked hard to develop geography as a 'real science' did not appreciate, or were unaware of, the philosophy of logical positivism which underlay their goal. Positivism has been described neatly as the 'hidden philosophy' of geography as a spatial science.

The spatial science school laid greater emphasis on the unity of science, on hypothetic-deductive methods and on the use of 'hard' data. Few of its members, however, were willing to accept the strict methods prescribed by the logical positivists as the only road to scientific knowledge. As the theoretical work of the spatial science school encouraged geographers to become more involved in planning, it became clear, however, that geography as a science is involved in shaping reality by explaining how situations are and what they could be like. Most research workers were perfectly aware of the political implications of this. Additionally, there was no agreement amongst spatial scientists as to the degree of universality of laws and models. Among the most quantitatively inclined geographers, few have been ready to argue that the laws and models of human geography and of the social sciences in general are unchanging and universal. Societies change and so do the laws of society – we can divide the world into different universes: capitalist, socialist, etc., with laws that differ between them.

There was also no common agreement on the role of science in advocating change. After the 1960s, geographers became more involved in **applied research**. Quantitative methods and models were, to a large extent, developed because they were thought to have considerable predictive value. This raises the immediate problem as to whether, in forecasting, we should rely on the projection of current trends or try to envisage alternative scenarios. The research workers' political outlook is very likely to affect their answers to that question. The present author considers that the important requirement of objectivity demands that we should analyse and state clearly those assumptions (or pseudo-constraints) that can change or may be changed, and also the consequences such changes may have on social development. It would be dishonest of research workers interested in the prediction of future changes not to do so.

Not all scientists will be able to attain such an ideal of objectivity. Those who support, even subconsciously, the maintenance of existing social structure soon come to believe that the overturning of certain assumptions or **pseudo-constraints** is unrealistic or wrong. Those who favour the overthrow of the social order, on the other hand, will easily overestimate the possibility of converting the rules of society into pseudo-constraints.

Gregory (1978, p. 77) maintained that 'the function of social science is to problematize what we conventionally regard as self-evident'. Gregory's argument leads on to **structuralist approaches** to human geography, whereas other critics of positivist tendencies within human geography adopted **humanistic approaches**.

HUMANISTIC APPROACHES

The discussion above implies differences between the **empirical-analytical** and the **historical-hermeneutic** sciences (see p. 108). These operate through different methodological frameworks: 'the former produces predictive knowledge for technical control, the latter interprets meanings to understand actors' (Taylor, 1985, p. 95). Following Bird (1993, p. 45) the humanistic approach may be characterized as follows (see also Box 4.3 and Figure 4.1):

- What exists is what people perceive to exist (its *ontological basis*).
- Knowledge is obtained subjectively in a world of meanings created by individuals (its *epistemological basis*).
- *Methodology* will then be interpretative, using methods which make it possible to investigate individual worlds and to emphasize individuality and subjectivity rather than replicability. Within humanistic approaches there are different opinions on methods, as there are various modifications of ontology and epistemology.

One important observation is that geography, more than any other natural or social science, is a visual science with similarities in this respect to architecture and the history of art. We analyse landscapes and provide methods of visual presentation in our teaching of geography. The visual models and thematic maps that explain geographical patterns increase our knowledge and understanding of the world, partly because they activate modes of thought that words and numbers cannot reach.

Neurosurgeons have shown that analytic, objective and so-called **scientific modes of thought** are associated with the left half of the brain (Figure 4.2). These thought processes, however, will only function when they have been stimulated by impulses from the artistic, holistic, visual, intuitive and irrational thoughts which are based in the right half of the brain. Visualization helps us to exploit the

Box 4.3   Ontology, epistemology and methodology

Figure 4.1   The epistemological funnel
Source: Translated from Löfgren, 1996, Figure 1.1

Ultimately any theoretical position (the outer frame of any research project) rests upon a basic philosophical position – an **ontology** (i.e. a theory of existence, an understanding of what really exists as opposed to what does not). This general ontology is also the frame for a more restricted **social ontology** which concerns the structure of a society, its elements and the basic mechanisms that link these. Giddens' structuration theory, Marxism and humanism are examples of social ontologies. Some of these also claim to be total ontologies; others not.

Although we might have a basic conception of our theoretical position, and might adhere to an ontology of how the world of existence and the social world are constituted, most of this existence is unknown to us. So we need a theory about how to get knowledge about the world – an **epistemology**, a theory of knowledge. Any research project or investigation is built upon a fundamental epistemological stance. It is the epistemology that guides the formulation of research problems, the evaluation of theory, the choice of appropriate techniques for empirical investigations and, above all, the interpretation of results. An ill-considered epistemological position can be an important cause of problems in a research project.

An epistemology is a general theory about how to look for knowledge as, for instance, exemplified in Sayer's (1984/1992) presentation of a realist epistemology (see Figure 4.6). To be applicable to research, an epistemology must be cast in a concrete form by way of models or programmes, i.e. **methodology**. The hypothetic-deductive approach is one such model. This gives certain rules about how research should, or ought to, be carried out if it is to be accepted as valid. If we stick to a specified methodology the findings we present could be replicated by others. If we do not use a specified methodology it is difficult for others to evaluate our results. Within each different methodology there is a host of different concrete **methods** we can apply. A method includes rules and recommendations (*shall*, *ought to* and *can*) on how to collect, work up, analyse and present data in relation to a given problem. A method is, in turn, built up from restricted and often interchangeable **techniques** on how to carry it out. For instance, collecting interview data can employ many different techniques within, for example, a qualitative research method. Finally, any techniques and methods used have to be adapted to the special research problem and context that are being investigated. Often **method triangulation** is needed: even though we may be starting from the same ontological and epistemological basis, there will be many different methods to choose from. The approach to a particular research project will often need to be a combination of different methods. Figure 4.1 represents research work as a funnel – from the basic ontology to practical research techniques.

Source: Adapted from Löfgren, 1996.

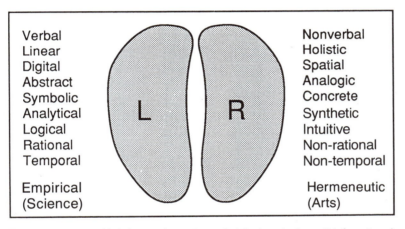

Figure 4.2   Comparison of left-hemisphere (L) and right-hemisphere (R) functional characteristics of the brain

full capacity of the brain but also implies that we must take into account those intuitive and less rational elements which contribute so much to our intellectual life. This goes to the core of humanistic approaches to geography in which, to a very large degree, research involves a personal approach employing both intuition and imaginative interpretation.

Granö suggests that the focal point of geography as an holistic science is neither the relation between humanity and nature as the object of study, nor the explanation of humanity's activity in terms of nature, nor the integration of humanity and nature in regions or landscapes. The focal point lies in the reciprocity between humanity's mind, which is the subject, and humanity's environment – an attempt to explain land and nature in terms of humanity: 'Man's perceptions, experience, knowledge and action form, together with his environment, a totality, a unity which constitutes the basic premise of geographical enquiry' (Granö, 1981, p. 23).

We cannot appreciate how the world, the 'real' environment, actually is: we can only know how we interpret it on the basis of our experience and knowledge. In Figure 4.3 experience is isolated from knowledge. The immanent sensation of the

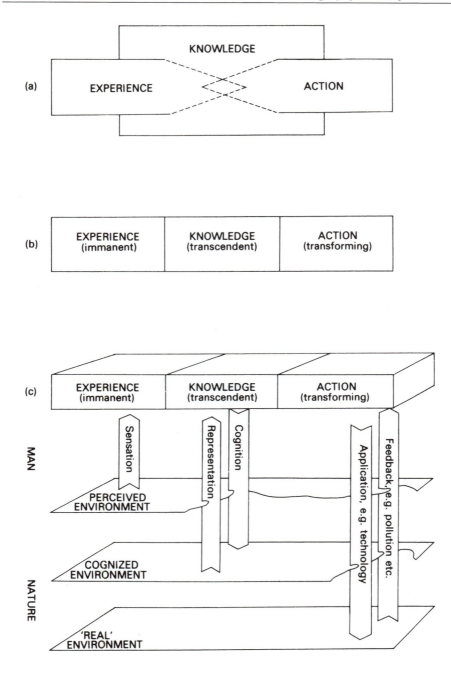

Figure 4.3   The cognized environment, which is the basis of our knowledge, is not identical to the 'real' environment in which our actions take place
Source: Granö, 1981, p. 24

environment, or **perceived environment**, is the same for everyone who is exposed to the sensation, but each person will have a different 'after-image', or **cognized environment**, based on his or her own experience, knowledge and/or memory. Our individually cognized environment constitutes the total of our knowledge of the environment, together with our imaginings of it and thoughts about it. These form the foundations of our actions, which, in turn, transform the **real environment**:

> As knowledge has developed, so the cognized environment created by mystical, speculative and subsequently rational thinking has been subjected to an increasing number of influences from perception. The development of empirical science meant that attempts were made to identify the cognized environment as nearly as possible with the real environment.
>
> (Granö, 1981 p. 25)

Humanists believe that this is impossible.

Humanistic approaches focus on the acting individual: the aim is to achieve an understanding of events, the thoughts underlying the actions that produced the world of experience: 'The role of the researcher is not that of a technician, one who promotes a certain solution, but rather that of a *provocateur*, one who promotes thought and reflection.' The goal is 'to increase *self-awareness* and *mutual awareness*' (Johnston, 1986b, p. 103).

Olsson (1978, p. 110) regards the spatial science and humanistic schools of geography as essentially opposites. The first aspires to formalized construction of knowledge, the latter to creative presentation. The humanistic approach cultivates the subjective interpretation of the world. Each one of us has his or her own subjective cognition of the 'landscapes of fear'. Tuan's book *Landscapes of Fear* (1980) represented his personal creative presentation of this phenomenon. Reports of such personal cognitions give us an understanding of the relations between the actor and the environment, but if we follow a **phenomenological** or an **existentialist** philosophy (see pp. 94–5) we cannot claim to reach general truth.

Working within the humanistic tradition, Guelke (1981; 1982) advocated **idealism** as an alternative to phenomenology and existentialism. Idealists use the method of sympathetic understanding (*verstehen*) and restrict themselves to thoughts that are in some ways rational in order to elucidate the reasons behind particular actions. The focus is on the actors, their intentions and meanings. This may be seen as an approach that supplements empirical research. Pickles (1985, pp. 52–3) points out that phenomenology is also not a type of irrational anti-science. **Phenomenology** provides the 'essential' basis for science, but does not replace it. In geography, humanistic aspects of the traditional schools, like the French regional school, have been clarified and developed further. There have also been some thought-provoking representations of, for instance, the feeling of home, identification with a place (Buttimer and Seamon, 1980), different conceptions of space (Sack, 1980), space, place and meaning (Jackson, 1989), the role of space and place in relation to gender (Duncan, 1996) and human territoriality (Malmberg, 1980; Sack, 1986).

There are two main lines of criticism of humanistic approaches. The first – positivist criticism – has, however, lost ground as the number of convinced, clear-cut positivist geographers diminishes rapidly. The second, largely structural critique, 'presents the atomistic focus on the individual in humanistic work as a distortion of reality: it gives individuals freedom to act when in fact they are very much constrained, if not constricted, by external circumstances over which they have little control' (Johnston, 1986a, p. 95).

STRUCTURALISM

Positivism implies that if 'we are to explain processes we must discover the regularities or universal laws governing their behaviour. Hence the thrust of research must be towards the discovery of order' (Sayer, 1985, p. 161). This strategy might work within the natural sciences, particularly in physics, but in human geography regularities tend to be approximate, temporally and spatially specific, and unique rather than repetitive. But as has been pointed out above, most models and theories put forward by the spatial science school are descriptive rather than explanatory. **Descriptive models**, like the **gravity model**, are valuable and scientifically respectable in so far as they provide new knowledge as to how things are. But science should provide explanations of the observed regularities; causes are not associated with correlations and regularities – these are **surface appearances**. We need to look for the mechanisms within **deep structures**.

The shapes of the pieces in a jigsaw puzzle or the forms in a cultural landscape do not tell us much about the machine or mechanism that cut the pieces. So we need a **structuralist approach** to obtain an understanding of how the mechanisms or driving forces within these structures (often called the **real level**) form outcomes at the **empirical level**.

In most forms of structuralism there are no deterministic relationships between the real and the empirical level. The French anthropologist Claude Levi-Strauss presented the analogy of a camshaft machine cutting jigsaw puzzles. The machine is designed to make only a certain number of movements, but these can cut a large variety of pieces. So we cannot understand the nature of the mechanism (the machine) if we only study the outcomes (the puzzles). We must study the mechanism itself but we cannot do this by direct observation. We must try to establish a theory of the mechanisms and find out whether this theory is consistent with appearances at the empirical level (Johnston, 1997, p. 215).

A variety of structural approaches has been identified within the humanities and social sciences. Linguists have established a school of structural linguistics that teaches there are certain imprinted, fundamental characteristics in language systems. Language systems contain certain basic principles (such as those of binary opposition) which are used to create meaning. The structural principles of language are so readily absorbed by children that they appear to be transmitted genetically. Structural linguistics analyses speech to identify these deep structures, which may be common to all languages.

Social anthropologists in particular Levi-Strauss, have transferred the ideas of structural linguistics to the study of social life, and they maintain that human behaviour is preordained by structural forces of which we are, in most cases, unaware. The myths and rites of a particular society may be studied as transformations of deep structures at the real level. Levi-Strauss argued, for example, that it is possible to identify certain basic elements within myths that may then be projected into a theory concerning the nature of these deep structures. Using the camshaft analogy given above, Levi-Strauss believes that we can understand the nature of its gear-box even if the only things we can actually observe are the characteristics of the jigsaw puzzles. Those types of structuralism that seek the universal, basic structures of the human mind have been called **structure as construct** types (Johnston, 1986a, p. 97). Levi-Strauss' ideas have played a role in geographical research through the works of Tuan, Sack and so on. One interesting example is concerned with children's perceptions of space and their acquisition of geographical knowledge (Sack, 1980). This form of structuralism must also, inevitably, be considered by anyone concerned with **semiotics** (see p. 134).

What Johnston (1986a, p. 101) calls **structure as process** types of structuralism have had a more direct impact on geography. They are not concerned with permanent mechanisms in deep structures or in the human mind, but investigate structures at the societal level. As societies change through dialectical processes, such structures are gradually transformed. Geographers are interested in the empirical effects of this process of structural change.

Part of the literature in this field may be regarded as **structural Marxism**. Marxism, as a theoretical base for social sciences, tries to understand how the mechanisms of the economic base are gradually changed in dialectical processes that occur between society and individual agents. In general, it is the mechanisms of the economic base that form the determining element; hence this approach may be called **quasi-determinist** (Johnston, 1986b, p. 56). The individuals or agents are more or less puppets who are manipulated by the economic mechanisms. The **ideological base** (religion, culture, etc.) is generally relegated to occupying a position of dependence on the **economic base** – if not directly determined by it. Duncan (1985, p. 176) maintains that

> the rhetoric of structural Marxism may sound radical, but the moral of human action is ultra-conservative, portraying actors as prisoners of fate caught in slowly grinding structures not of their own making, waiting for the evolution of the social structures through contradictions to move them on to a higher stage in the evolutionary process.

Moreover, as **historical materialism** is concerned with hidden structures, the theories can never be finally demonstrated or empirically verified: 'The "test" of such theories is provided by the success or failure of actions founded upon them' (Johnston, 1986a, p. 109). So Marxism is as much a guide to political practice as a scientific method.

Harvey (1973, pp. 129–30) stated that the essential difference between positivism and Marxism is 'that positivism simply seeks to understand the world whereas Marxism seeks to change it'. The political aims of Marxist science are also expressed by Richard Peet (1977, p. 21):

> Marxist science begins with a material analysis of society, proceeds through a critique of capitalist control of the material base of society, and proposes solutions in terms of social ownership of that material (economic) base . . . the political aims of Marxism provide a common scientific purpose. As a holistic, revolutionary science, Marxism provides a firm theoretical base for the radical movement in geography.

Geographical studies inspired by structural Marxism are mainly found within economic geography (see, e.g. Massey, 1984; Thrift, 1986) and social geography

(Harvey, 1973; 1985a; 1985b). Bennett (1985, p. 221) criticizes Harvey for regarding the effects on the physical environment as merely a product of the social system, with society the result of social control. Resource problems are seen (*ibid.*) only as distributive problems deriving from class structure:

> This ignoring of physical questions is sometimes justified by the view that the concept of 'human agency' can be applied only to 'human matters'; if this is so then this justification accepts social theory as at best a partial theory and hence of limited relevance to geography which has physical–human relations at its core.

Quaini (1982) demonstrates, however, some interesting relationships between Marxism and ecological questions, and Marxist-inspired social ecology has played an important role in some schools of geography (see, for instance, Peet and Watts, eds., 1996).

In Figure 4.4, Chorley (1976) shows a dichotomous classification of positivism and 'behaviouralism' (encompassing what we earlier have presented as humanistic geography, behavioural geography and radical geography or structural Marxism). In the 1960s, as Bird (1993, p. 48) points out, there was a movement from the right to the left across the dichotomous divide; human geography adopted the positivist methodologies of 'hard science' and became spatial science.

Since the 1970s there has been a drift away from **positivism**, but the purer forms of physical geography and parts of human geography, particularly within locational studies, remain happily positivistic. This part of geography has also increasingly become involved in debates on the philosophy of science, as has been neatly demonstrated by Haines-Young and Petch (1986). In physical geography they find, however, that a better understanding of research problems can be achieved within the positivist traditions, Popper being a good guide. On the other hand, in the 1980s and 1990s, the move to 'soft systems' led to new trends in human geography. Among these we will now consider **structuration theory**, **realism**, **poststructuralism** and **postmodernism**.

STRUCTURATION THEORY

In the 1980s, many observers regarded human geography as caught between the determinism of structural Marxism (which tries to 'read off' the specifics of place from general laws of capitalism) and the voluntarism of most humanistic geography (in which events result from purely individual intentions). In the tensions between these positions some human geographers sought a resolution in 'a model which allows autonomy to social consciousness within a context determined, in final analysis, by social being, one in which history is neither willed nor fortuitous, and neither lawed nor illogical' (Peet, 1996, p. 873).

To achieve this resolution requires theories, and also analytic mediations. **Structuration theories**, as presented among others by Bourdieu (1977), Bhaskar (1979) and Giddens (1979; 1984), aimed at such a resolution. Among these authors it is the work of the British sociologist Anthony Giddens that has had the most profound influence on human geography. His theories were founded on a critique of the 'structural and interpretative sociologies that were connected to, but not synonymous with, the Marxist and humanist traditions that surfaced in the human geography of the early 1970s' (Cloke *et al.*, 1991, p. 93).

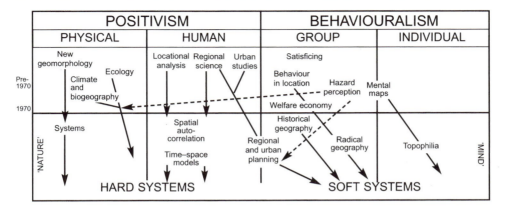

Figure 4.4 Geographical studies arranged according to relative 'hardness' or 'softness'
Source: This is an extract redrawn with permission from a diagram without a legend in Chorley (1976, p. 31)

We need to repeat here that any theoretical position rests ultimately upon two philosophical components: **ontology** and **epistemology** (see Box 4.3 p. 118). Epistemology gives guidance on how to work scientifically; ontology provides a basis for understanding the world. **Giddens' structuration theory** is largely ontological in its orientation. Giddens (1984) explains that he has tried to develop an ontology of human society (concentrating on how to theorize **human agency**) and to consider the implications of this theorizing for the analysis of social institutions. 'Human agency' refers to people's capabilities and their related activities or 'behaviour', but not to the agents themselves.

In **structural Marxism**, agents are ruled by structures. Giddens understands structures as being created and re-created through human agency; the agents' position is central. The individual agent has only an incomplete knowledge of either the empirical world or the 'mechanisms' (social rules) of society that structure his or her actions. If we are to make a scientific analysis, however, we must assume the agent's actions to be rational within the context of his or her incomplete knowledge. In Giddens' model, the **agents** are not puppets – they are always conscious and unconscious interpreters between the real level and the empirical level. The agents interpret and transform the empirical world, but these interpretations and transformations are constrained and enabled by the structures at the real level. Even the structures of the economic and ideological base are interpreted and transformed by the agents.

Individuals are born into societies that entrap them within social structures, which both constrain and also enable them. While we live out our lives under the rules made by society, we are also reproducing these rules, but not necessarily in the same form. Since every individual is an agent, the rules are gradually changed. The gradual transformation of a language is a good example of this process. Religions and political ideologies are also gradually transformed through individual practice.

At the core of Giddens' theory lie the concepts of **'structure'**, **'action'**, **'social system'**, **'social practice'**, **'context'** and **'duality of structure'**. Structures are sets of rules (constraints) and resources (capacities or possibilities) which 'exist only as memory traces, the organic basis of human knowledgeability, and as instantiated in action' (Giddens, 1984, p. 377). Structures should be understood as rules and resources for action. There are rules for the constitution of meaning and rules for the sanctioning of social conduct. These rules can be either informal – as social norms – or formal – as laws and bureaucratic regulations.

Social rules are implemented through the interaction of people with each other. Social rules also structure interaction, and the rules which structure interaction are themselves reproduced by the process of interaction. We follow such rules consciously or unconsciously, depending on the degree to which they are self-evident to us. **Resources** may be regarded as either **allocative** or **authoritative**. Allocative resources control material relations (the economy) both in the private and public sphere. Authoritative resources control people both privately (domestic abuse) and in public (politics). It follows from this that 'the structural properties of social systems, as formulated in the rules and resources postulated by Giddens, are both a medium of social practices and an end result of social practices' (Cloke *et al.*, 1991, p. 102).

For Giddens, the concept 'action' represents the routine actions of daily life. Action is regarded as a continuous process rather than as a series of isolated single actions with specific intentions or aims. Giddens does not deny that such isolated, intentional actions exist, but argues that most intentions and aims develop in a continuous, reflexive process, in which we think through our own actions or the actions of others. He calls this process 'the reflexive monitoring of action' (Giddens, 1984, p. 376). The knowledge upon which our actions are based includes both **discursive consciousness** (the types of knowledge we can explain and discuss in words) and **practical consciousness** (knowledge we see as more or less self-evident, which we use in our daily life, and which we cannot necessarily explain in words). Routine and repetitive tasks, which we do without thinking how or why, are based on this kind of knowledge.

However skilled and competent **agents** may be, their knowledge is limited. As a result, conditions of action may remain unacknowledged and the consequences of action may be unintended. Agents are just as much influenced by the actions of others as they are by the rules and limitations of structures.

Structuration theory thus takes account of both the acting subject and the social structures that exist independent of human consciousness, but not independent of human practice. Structures set the conditions for human actions, but they are also the results of human actions. Individuals are formed by society and its institutions, but they are also skilled agents who direct their own lives through actions (agency). This is called the **duality of structure**. Cloke *et al.* (1991, p. 98) point out that the notion of duality represents a key stage in Giddens' structuration journey. We need

1. to recognise the duality of structure: that is, the manner in which structures enable behaviour, but behaviour can eventually influence and reconstitute structure; and
2. to recognise the duality of structure and agency: that is, to transcend the dualism of deterministic views of structure and voluntaristic views of agency.

Action and structure are theoretical concepts which, in real life, are knitted together by social practice. Repeated social practices, in the form of social relations and routines, create a social system. Well established social systems create social institutions, such as schools and workplaces.

Giddens (1984) considers that social actions always take place within a framework – an empirical connection. From a geographical point of view, Giddens' suggests that links between structuration theory and Hägerstrand's **time-space geography** are important (see pp. 81–82). Social systems are not only structured by rules and resources; they are also structured in time and space. Time and space are not passive frames for action but part of the action itself. Giddens uses the term '**locale**', which many geographers have (somewhat wrongly) understood as '**place**'. Actions are imprinted by socialization in 'locales', which are created and re-created by people who use their intellectual capacities within the frames set by social rules and mechanisms. While empirical research may show us how this happens and describes what constitutes the mosaic of a geographical pattern, only theoretical research can answer the more fundamental 'why' questions about the relations between agents and what they regard as constraining mechanisms.

From a geographical point of view, Giddens' discussion of the spatial concept is fairly simple. He has been criticized for not knowing enough about the theoretical discussions among geographers concerning the continuous production and reproduction of space and on the different concepts of space. Geographers have, however, used structuration theory in their analyses of space. For instance, Pred (1984) has identified 'place' as having been produced in a 'historically contingent process', something which is changed continually as a result of the social actions of the people who have lived there, and which influences the same people and their activities at the same time.

Structuration theory has been criticized for not giving direct guidance on how to proceed in scientific investigations, and for the absence of formal links between structuration theory and empirical studies. **Realism** may be seen as an attempt to achieve this linkage.

REALISM

The principles of **transcendental idealism** (as promoted by Immanuel Kant and subsequent philosophers, including **phenomenologists**) claim that knowledge can only be produced through – and therefore only consists of – our ideas, perceptions and interpretations. In this philosophy, the natural world becomes a construction of the human mind, or a construction of the scientific community. Bhaskar (1978) compares this with **classical empiricism** (sometimes called **naive realism**, and connected with **positivism**), where regularities are discovered from observations of the physical and social world around us. Scientific knowledge is hence based on direct responses to the stimulus of given empirical facts. As promoted by Bhaskar (*ibid.*), **realism** (in the form of **transcendental realism**) may be regarded as occupying a position between classical empiricism and transcendental idealism. The word 'transcendental' may seem obscure, but it refers to any philosophy based upon the doctrine that the principles of reality can be discovered by studying the processes of thought. This concept is in some way a concession to Kant, who proposed that knowledge is based upon certain a priori elements of experience (see p. 24). However, if we combine Kant's basic ideas

	Domain of real	Domain of actual	Domain of empirical
Mechanisms	√		
Events	√	√	
Experiences	√	√	√

Figure 4.5   The three domains of reality
Source: Bhaskar, 1975, p. 56; Cloke *et al.*, 1991, p. 137

with realism we achieve a philosophy that proposes there is a 'real' world of physical things, which exist whether or not we perceive or recognize them. **Transcendental realists** 'use terms like "practical adequacy" for corroboration, and believe that phenomena in the real world are structured and possess causal powers and liabilities' (Bird, 1993, p. 110).

The **ontological basis** for transcendental realism as developed by Bhaskar (1975), suggests that 'yes there is a *real* world', but that some of its most significant components are not immediately observable. There are two types of such non-observable structures. The first are in the form of capitalist or other structures and forces of power, and theoretical (structuralist) analysis may elucidate how these work. The others are structures that are found in the form of experiences and conceptions 'in people's heads', which cannot be observed or measured directly but which are the basis for actions and, in this way, shape events in society. We may here see connections to both Giddens' structuration theory and humanism. It is recognized that knowledge about certain phenomena can only be achieved through the study of human conceptions. Human practices cannot be understood only as rational and functional, but must also *work meaningfully* for the people involved. This is neatly illustrated by Aase (1999) in Box 4.4 (pp. 135–36). Bhaskar (1975, p. 56) recognizes three distinct domains of reality (Figure 4.5), which tend to be oversimplified and conflated in positivist (empiricist) approaches:

1) The *real* – mechanisms (which are unobservable).
2) The *actual* – events (which are observable phenomena).
3) The *empirical* – experiences of events.

Bhaskar (1979) questions whether nature and society can be studied in the same way, and he maintains that transcendental realism opens up a unity of scientific method in terms of the form taken by social and natural scientific knowledge, the reasoning by which it is produced, and the concepts which theoretize its production. In both cases, the objects of scientific inquiry are real structures, and the method of science implies moving from knowledge of manifest phenomena to the structures that generate them (Peet, 1998, p. 168).

So far, our discussion has broadly underpinned the **ontological** basis of realism. The real contribution to a **realist epistemology** for geographical research has been provided by Andrew Sayer in his book *Method in Social Science: A Realist Approach* (1984; new edition 1992).

Positivism may suit the analysis of **closed systems**. As Sayer (*ibid.*, p. 122) points out, in the objects of study in the natural sciences, 'closed systems may exist naturally (e.g. the solar system) or may be produced in experiments or machines'. For such systems it is possible to make universal statements and to conceptualize a constant mechanism that causes regularities. It is also possible to formulate a set of constant conditions in which the causal mechanism operates. An initial premise for Sayer (*ibid.*, p. 123) is, however, that

> the social sciences deal with open systems but lack the advantage of their equivalents in natural science of having relevant system sciences on which to draw. One of the main reasons for the openness of social systems is the fact that we can interpret the same material condition and statements in *different* ways and hence learn new ways of responding, so that effectively we become different kinds of people.

Open social systems are thus much more complex to analyse: more than one mechanism may explain an event, and mechanisms may change during our investigation, and even possibly because of it. In open systems, changing external conditions make it impossible to extract regularities and explanations from actual events. Theory can elucidate the **necessary conditions** that characterize the mechanisms acting to create specified actual events, but these are not **sufficient conditions** for an explanation.

Sayer (*ibid.*, p. 92) points out the importance of asking qualitative questions about the nature of our objects with the aim of identifying social **structures**. We may, for instance, observe that some industries are more strike prone than others. Many social scientists will try to find causes for this through statistical analysis of possibly independent variables (such as size of establishment, union membership or gender composition of staff). We might find a correlation between size and strike-proneness, but is it size itself, or social relations, or forms of management systems associated with different sizes of firms that are important? New statistical tests cannot solve this problem. To answer the qualitative questions about the nature of our objects, science must **abstract** from the many conditions of an object to focus on those it is believed have a significant effect. To abstract means to isolate in thought a one-sided or partial aspect of a concrete research object, and to form abstract concepts about it before going back to the concrete armed with a host of theoretical understandings (Peet, 1998, p. 168). Sayer (1992, p. 92) maintains that there is a continuum rather than a dichotomy between abstract and concrete. However, abstraction is necessary to distinguish essential characteristics from the maze of incidental characteristics:

> Abstraction is particularly important for the identification of **structures**. These can be defined as sets of internally related objects or practices. The landlord–tenant relation itself presupposes the existence of private property, rent, the production of an economic surplus and so on; together they form a structure. Contrary to common assumption, structures include not only big social objects such as the international division of labour, but small ones at the interpersonal and personal levels (e.g. conceptual structures) and still smaller non-social ones at the neurological level and beyond. Within social structures there are particular 'positions' associated with certain roles. It is particularly important to distinguish the occupant of a position from the position itself.

It is often assumed that **individuals** have free will and must be held responsible for their actions. However, people are also constrained by rules and structures, even though such structures only exist where people reproduce them (see Giddens' structuration theory above). Structures are thereby seen as durable and sometimes capable of causing social change. However, they can also lock their occupants into particular role positions. Durable social structures have positions associated with roles independent of the persons occupying them. As Bhaskar (1979, p. 44) states: 'People do not marry to reproduce nuclear family or work to reproduce capitalist economy. Yet it is nevertheless the unintended consequence of, as it is also a necessary condition for, their activity.'

It is thus important to distinguish between **necessary causal powers (or conditions)** and **contingent conditions**, and also between **internal** and **external relations**. Gunpowder possesses the necessary conditions to explode, but whether it will explode or not depends on an 'accidental' spark. In most cases an industrial workforce has the necessary power to go on strike; whether they strike or not depends upon particular contingent conditions, which may be highly local and on which generalization is impossible. As demonstrated by the miners' strike described in Box 1.3, p. 9, necessary causal powers operate only at a high, general (in this case, national) level, whereas the local level is associated with external, contingent relations. Sayer (1991, p. 300) points out, however, that

> the 'general-causes-and-local-contingent-effects' model may sometimes be appropriate, but realism gives no special privilege to it, as many commentators on locality research (both advocates and opponents) have supposed. Other variants, such as local causes and local effects, or local causes and general or global effects, or global causes and global effects, are just as compatible with realism, and just as relevant for understanding society.

The relation between you and a clothing shop is **external** in the sense that either you or the shop can exist without the other. If you buy your clothes in that shop, however, the relationship becomes **contingent**. In contrast, the relationship between landlord and tenant is **internal** and **necessary** since the existence of either presupposes the existence of the other. How the landlord treats the tenant is, however, **contingent** on (for instance) the personalities of the persons, whether they are male or female, etc.

When striving to put realism into practice we need theoretical categories to understand theoretical relations. For this reason it is one of Sayer's main intentions to clarify obscure theoretical concepts. The realist terminology for basic theoretical categories of necessary relations is called **abstract theory**. In this sense 'abstract' does not mean 'vague' but refers to a particular relationship between causal powers and an object of study. Abstract theoretical research deals with structures and mechanisms. Events are only considered as possible outcomes: 'Abstract theory analyses objects in terms of their constitutive structures, as parts of wider structures and in terms of their causal power. Concrete research looks at what happens when these combine' (Sayer, 1984/1992, p. 107; Figure 4.6).

Concrete research is the empirical study of events and not-necessary (contingent) relations. Such relations represent all the conditions relating to the individuals, firms, organizations or whatever we might study, which are not part of the actual internal relations. At the abstract level, we find a number of structures with causal power that can be activated through certain mechanisms. The results can be studied at the concrete level of events. The same mechanisms can produce different events, as indicated by the vertical lines in Figure 4.6, and the same event

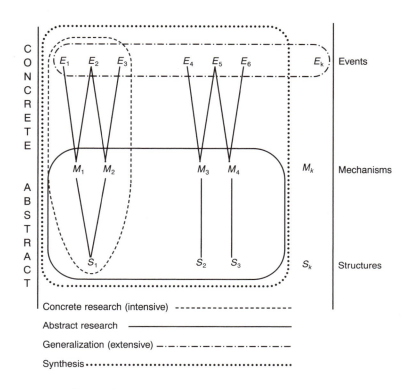

Figure 4.6   Types of research
Source: From Sayer, 1984, p. 215; Cloke *et al.*, 1991, p. 153

may have different causes. Concrete research focuses basically on the events, but Sayer recommends us to employ a combination of concrete and abstract research.

In Figure 4.6 **generalization (extensive research)** may illustrate the type of concrete research activity within spatial science that tries to find explanations by searching for regularities in extensive quantitative empirical investigations. Sayer recommends **intensive concrete research** that, on the basis of abstract considerations of some structures and mechanisms, analyses their possible effects in limited empirical **case studies** to achieve an understanding of the functions of necessary and contingent relations.

For instance, Bjørlykke (1996) compared the establishment of oil terminals in Shetland, UK with Øygarden, Norway. She assumed that the oil companies would be embedded in the rationality (necessary relations) of the global market economy and that the outcomes (events) in both locations would be very much alike. Intensive concrete research, based on qualitative methods, revealed, however, that contingent relations at the local level led to events that were rather different at the two locations. So with different contingent conditions, the same mechanism may invoke different events; on the other hand, the same kind of event may have different causes. It is important to note here that the form of contingent relations must always remain an empirical question, answered through observation of the actual cases. However, structural relations might be found at different geographical levels. The international relations of oil companies might be regarded as being at a structural level, but the local rules of the Shetland community could equally be regarded as structural. Depending on the level of the investigation, these might be seen as either necessary or contingent factors.

Some would like to include a fourth level in the vertical dimension of Figure 4.6 to cover **agency** (in the form of meanings, experiences, interpretations, etc.). But as agency can form structures, can function as causes or be considered as events, Sayer (1992) maintains that agency is already included. This means it is impossible to equate the concepts of structure and mechanism in Figure 4.6 with structure and agency in structuration theory. Realist discussions on structure and agency include, in general, a series of warnings about the complexity of these issues, particularly of course because realism concerns itself with establishing a guide to research. **Epistemology** – the theory of knowledge – is at the heart of realism, whereas structuration is mainly concerned with **ontology** – the theory of existence.

A SIMPLIFIED MODEL OF RELATIONS BETWEEN THE EMPIRICAL, AGENTS AND STRUCTURES

When geographers observe the geographical mosaic, they can, says Johnston (1985, p. 335), regard every place as *unique* but not as *singular*. General rules and societal structures exist, but they are adapted at each place in a unique combination of individual interpretations made by the agents at the specific 'locale'. The structural rules of society, like the capitalist mode of production, are a given. Capitalism is not, however, a deterministic mechanism; its imperatives have been interpreted in many ways. Former interpretations are imprinted as a part of culture and create frames for new interpretations. Actions based upon these interpretations also result in distinct physical landscapes and social patterns, which in turn constrain and enable new interpretations and actions (see Figure 4.7).

We have argued elsewhere (Holt-Jensen, 1988, p. 137) that it might also be possible to regard geography's traditional role as a bridge-builder between the natural and social sciences in these terms. While social scientists are primarily interested in the empirical results of interactions between human beings, geographers are additionally concerned with the natural environment and with the constraints and relations between this environment and humanity's activities. The limitations and possibilities of the physical environment are interpreted by **agents** in just the same way as the limits and possibilities set by both (hidden) social mechanisms and empirical social patterns. It is of crucial interest to society how and why environmental constraints are interpreted so variously by different agents at the same place, and how interpretations and intended actions *vis-à-vis* the physical environment can differ from place to place. Places matter and so does geography as a mediator and bridge-builder between different approaches.

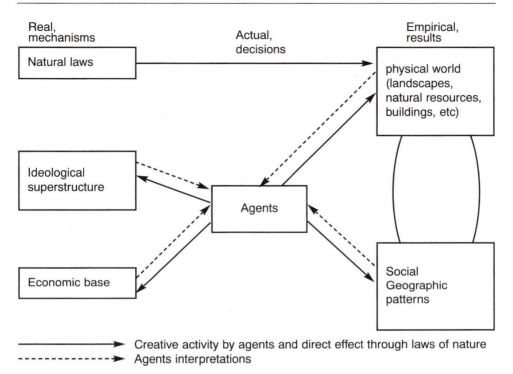

Creative activity by agents and direct effect through laws of nature
Agents interpretations

Figure 4.7 A simplified model of relations. Social geographic patterns include demographic structure, political divisions, patterns of welfare etc.

POSTMODERNISM AND POSTSTRUCTURALISM

Structuralism, structuration and realism are philosophies that somehow believe in **modernity**, particularly in the potential of modern science to create evidence, fact and truth that will set us free. However, a revival of interest in anti-modern philosophies (Nietzsche and Heidegger particularly) gradually grew within the radical, social-anarchist movement that was a product of the 1968 upheavals among students and working-class militants in Paris. Notions of a **postindustrial society** and critiques of modernist, western models for economic development in Third World countries merged with cultural critiques of modern architecture and aesthetics. Power relations in our postindustrial society have changed, making an orthodox Marxist theory of power obsolete. Power no longer resides in institutions or in the economy but resides instead in codes, simulations and the media, and it becomes increasingly abstract, maintains Baudrillard (1987). As opposed to Marx's productivism, **poststructural** theory emphasizes consumption, in which (for instance) the broadcast media, particularly television, create a hyper-reality that is more than real (Peet, 1998, p. 213). **Poststructuralism** generally interprets modern society as a system of power, and it expresses a fundamental scepticism towards any form of totalitarian regime – capitalist as well as communist. This has led to new interest in **social anarchism** (see pp. 35–36).

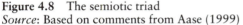

Figure 4.8 The semiotic triad
Source: Based on comments from Aase (1999)

There is no sharp difference between poststructural and postmodern philosophies. Generally, however, *poststructural philosophy* criticizes the certainties of modern knowledge, as with its claim to coherence, neutrality and truth, while *postmodern philosophy* carries this further into an alternative discourse based on oppositional modes of understanding. Postmodernism proposes new ways of being a person which involve, on the one hand the liberation of desire, yet on the other, utter cynicism or nihilistic 'resistance' to the forces of modernity.

(Peet, 1998, p. 208)

We may state, however, that **poststructuralism** basically rests upon a philosophical position that relates both to ontology (the theory of existence) and epistemology (the theory of knowledge). On the other hand, **postmodernism** relates more directly to factual changes in our societies, and actions and reactions to those changes. Poststructuralism has a direct bearing on scientific methods. If we adopt a poststructuralist stance, we must take the process of 'writing' human geography – of representing our findings in words, sounds and pictures – far more seriously. This must involve a critical reconsideration of our research methods. Box 4.2 (p. 112) shows how this leads on to **self-reflection**, i.e. reflection on your own role as a researcher. Human geographers need better training in interpretation and reflection, and on the role of concepts and signs in the research process. '**Semiotics**' (a Greek word that means 'showing' or 'signifying') encourages research into **signs** and meanings, and has succeeded in expanding our understanding in many recent research projects. Semiotics attempts to uncover the underlying cultural codes that determine how people interpret the spoken word and other forms of human communications. The **semiotic triad** (Figure 4.8) provides a background for the analysis of the relations between an object and its interpreters. For instance, anyone investigating a proposal to build a nuclear power station in an area of outstanding natural beauty would have to communicate with both the officials of the energy company and the local inhabitants. These interest groups are called the **interpretants**: interpretants interpret the symbols and other signs used, for instance, in the description of the area and the power project. In the communication between the different groups and the investigator, other signs might be conveyed, which are interpreted by the investigator. There are three main types of **signs**: symbols, icons and indexes. All words are symbols. An icon is a 'motivated' sign that looks like the object (for instance, a map of the site). An index is a motivated sign that links to the object in the same way as smoke is associated with fire (or nuclear power with radioactive waste). The concept **metaphor** is often used to describe a kind of iconic sign by which, for instance, a new phenomenon is given meaning by comparing it as, or akin to, something already well known.

Box 4.4   The use of metaphor in Himalayan resource management

In the Sai Valley in northern Pakistan, a hydroelectric power project created the administrative problem of adapting traditional practices in land and water rights to formal state rules. Adaptation is not only a question of finding the practices that seem objectively to function best but it is also a matter of finding practices that *work meaningfully* for local people.

A hydroelectric power project is a rather dramatic example of modernization that might lead to the fragmentation of the local community if not handled wisely. In such a process people may try to create a feeling of 'integrity' through their use of **metaphor**. A metaphor is a word or phrase that is applied to an object or concept it does not literally denote in order to suggest a comparison with the other object or concept (as in 'a mighty fortress is our God'). In this connection it implies, however, that meanings from known realms of life are ascribed to new, modern phenomena in order to restore a notion of integrated reality.

> In the case of the Sai Valley development, the inhabitants searched in their culture and history for meaning that could be ascribed – transferred, so to speak – to the new arrangements. In this process, not only new practices are assumed, but the practices are loaded with local, experience-near meaning. The cultural change that is usually read into events of modernisation is exactly the result of such metaphorisation.
>
> (Aase, 1999, forthcoming)

Precipitation in the region's valleys is, for agriculture, insufficient but glacial meltwater from the mountains and cultivable land is plentiful. For centuries, the construction of long irrigation channels in the steep mountain sides provided the necessary extensions to agricultural land, and increased animal husbandry provided manure to fertilize the extra growing areas. Land and water rights depended on a local power balance between the villages, and conflicts were solved in traditional ways. Today, vital resources are managed formally by the Pakistani state. In the case of the new power station, the authorities, wisely, tried to reach a consensus with the local villages about the ownership of the land that became available for irrigation.

To utilize the height difference for the production of power, it was proposed that the bulk of the water from the Sai River should be diverted to the village of Pari in the deeper Gilgit River valley (see Figure 4.9). A new agricultural area, Pari Das, could thus be established close to Pari. This, combined with the fact that the villages along Sai are divided from Pari Das by a high mountain ridge, would make it natural enough to give

Figure 4.9   Three-dimensional diagram of the region. The dotted line shows the diversion channel for water from the Sai River to Pari
Source: From Aase 1999 forthcoming

the new land to the villagers of Pari. But, astonishingly, Chakker Kot village in the Sai Valley emerged as the winner of the new land, and this was readily accepted by the villagers of Pari. The new owners have to climb the mountain ridge to reach their new land, which is in an area that was previously hardly ever used by the villagers for either pasturing or forestry!

Which arguments were applied to establish legitimacy to this land, and how was consensus reached? The answer is rather strange since it relates to a division between the villages in portering duties (the duty to carry goods) the villages held in the period between the 1880s and 1947 for the Kashmiri garrison, which was located further upstream above Pari at Gilgit. Even though Chakker Kot is not located in the Indus–Gilgit Valleys (which the porter route followed), its inhabitants were obliged to carry goods between the confluence of the Gilgit and Indus Rivers to Pari. As the new land is located a little downstream from Pari, Chakker Kot argued that the land was theirs since it was situated in the area of their old portering duty. This was accepted by the neighbouring villages. Since the Pari villagers also accepted this, we must assume that the episode played on strings of common meaning in the villages. However, this common meaning must, notes Aase, be related to another field of experience, where a similar constellation of meanings is accepted and practised. We cannot follow the arguments further here, but we do need to point out that Aase documents that the land-rights claims were accepted because of their metaphorical resemblance to traditional channel-maintenance arrangements, and so to rules connected to individual farmers' land rights.

Chakker Kot villagers, who gained ownership of the new land, eagerly support the power scheme, whereas villages further down the Sai River strongly oppose it. In this discourse, the upstream and downstream settlements refer to differing systems of meaning – namely, to tradition and religion. So modernization is not just the assimilation of traditional society into a universalistic state. Rather, people try to make sense of the changes by searching for meaning in tradition, religion and present practices: 'Out of this innovative, cultural process, new practices emerge, combining meaning from tradition with the challenges of modernity . . . Thus, the Sai case is an expressive example of local particularism in a globalizing world' (*ibid.*).

An approach based on semiotics is needed, in particular, when we carry out research into foreign cultures. It will often be impossible to understand factual events if we do not achieve an intersubjective understanding of the **meaning** embedded in a local culture or even in an individual creed. A study carried out by Aase (1999) in the Sai Valley in northern Pakistan illustrates this point (see Box 4.4).

Michel Foucault (1926–84), professor of the history of systems of thought at the College of France, has been regarded as the main proponent of **postmodernist philosophy**. Throughout his work he maintains a hostile position towards modernity, for what he finds to be its repressive, totalizing modes of thought. He favours 'difference, fragmentation, and discontinuity, multiple forms of analysis rather than single truths, microanalyses interwoven with macroanalyses, pure opinion mixed with exact historical studies' (Peet, 1998, p. 200). He argues against global, totalitarian theories such as Freud's theories of psychoanalysis and Marxism, favouring instead a non-centralized, autonomous, local theoretical production whose validity does not depend on approval from established regimes of thought. In practical terms this endorses a 'bottom-up' approach to research – a focus on **everyday life** and local perceptions.

Postmodernism is typically sceptical about 'grand theories' and established intellectual creeds such as Marxism, humanism, realism or structuration theory. Postmodernist thought suggests that we need to observe humanity from a humbler and more eclectic position. Another significant aspect of postmodernist thought is its focus on **difference** – an alertness to and appreciation of the many differences that exist among the people of the world.

The term **postmodernist** has been applied to many cultural objects, including paintings, novels, plays and films. Postmodernist trends are also visible in architecture and town planning. The schools of modernist and '**placeless**' architecture represented by Le Corbusier and functionalism are no longer fashionable: grand planning designs are out of favour and strict spatial differentiations of functions (housing, parks, services, industry, etc.) are no longer countenanced. We now look for variety in townscapes – mixtures of functions and the rehabilitation of old buildings rather than the erection of new, modern, high-rise and functional blocks. Instead of expert, top-down planning we now emphasize participation and the 'bottom-up' involvement of citizens and interest groups in the planning process.

Much of the geographical interest in postmodernity has been focused on the economies of western societies, where traditional capitalism is changing. Lash and Urry, in their book *The End of Organised Capitalism* (1987), analyse the restless and essentially 'disorganized' character of contemporary capitalism. Harvey (1989) sketches the impact of a new flexible capitalism on the revised experiences of time and space upon which postmodern cultural developments crucially depend. In societies where most people find employment in tertiary and quarternary activities and only a minority work in agriculture and manufacturing, postmodern developments have influenced the tastes and lifestyles of a new middle class which reflects the recent restructuring of the capitalist system. With Harvey, however, I still believe that analysis and theoretical understanding provide a map for radical political action as part of a liberatory politics. Harvey finds this politics grounded in **historical materialism**, but with important differences from orthodox Marxism (Peet, 1998, p. 221).

Foucault (1986) pointed out that 'the present epoch will perhaps be above all the epoch of space'. Soja (1989), partly basing his argument on Foucault, associates 'modern' with history and 'postmodern' with geography, and finds a new, postmodern, critical geography emerging, 'which reasserts the significance of space or, rather, sees time and space together in a creative commingling as the vertical and horizontal dimensions of being' (Peet, 1998, p. 223).

Postmodernism warns us against grand theories and urges us to employ reflection and self-reflection at all stages of a research process. Sayer (1993) points out, however, that the postmodernist preoccupation with the problems of discourse and language, its suspicion of concepts of truth and falsity and of empirical testing, mean it is in danger of flipping over into an equally untenable position of cultural relativism: 'Moreover, to point out the irreducibly metaphorical and rhetorical character of knowledge does not mean that logic, protocols of reason, just disappear. They remain, indispensable as ever: no flip from reason to rhetoric is needed' (*ibid.*). Sayer argues that, instead of flipping from naive positivism to postmodernist relativism, **realism** offers a third way between them. We may drop the idea of truth as absolute but we cannot ignore the relationship between discourse and the world.

Dear (1988) argues that postmodernism makes geography an offer it cannot afford to refuse because the positioning of space is the privileged realm of postmodernism. But as postmodernism holds there cannot be any grand theory of human geography, we might be concerned with the extremes of deconstructive relativism: 'It is from this paradoxical position, enticed by postmodernism, worried by its anarchy, remaining modern, that Dear wants to reconstruct human geography' (Peet, 1998, p. 218).

In addition, says Gregory (1989, p. 92), when reflecting on 'areal differentiation and postmodern human geography'

> we need in part, to go *back* to the question of areal differentiation: but armed with a new theoretical sensitivity towards the world in which we live and the ways in which we represent it. Whether we focus on 'order' or 'disorder' or on the tension between the two – and no matter how we choose to define these terms – we still have to 'look'. We are still making geography.

We are still making geography, but if we now step aside and look at the millennium-long history of the discipline, we realize that in the last decades academic geography has developed into a theoretical discipline that is not readily understood by the ordinary citizen. To succeed in academic geography, a student needs years of professional training. Geography clearly has academic and theoretical obligations in the world of learning, but does it also have other obligations, obligations in the general education of the population at large? Before looking in Chapter 5 at the discipline's central theoretical concepts, we need to consider the discipline's different roles and obligations. And if we find that other roles rather than the academic are of importance, should this have a bearing on how we teach and define the discipline at universities?

TO UNITE A VERNACULAR AND AN ACADEMIC DEFINITION OF GEOGRAPHY

As has been pointed out by Asheim (1990a, p. 290), Johnston (1986c) makes a distinction between a **vernacular** and an **academic** definition of geography. Asheim would prefer the discipline to develop along the lines of its purely academic definition. This would mean a discipline that is constantly changing and developing in conjunction with research problems, philosophies and methods that academic researchers currently regarded as interesting and rewarding. Such a stance would give the academic full freedom to take up any kind of research theme, with no regard for the history, traditions or public expectations of geography as a discipline. The present author cannot support such a purely academic definition of geography. He believes we have to regard the vernacular definition (the public expectations of our discipline, its history and traditions) as an important reason for its existence. Academic disciplines can be compared to political parties. Both compete in public markets: political parties compete with each other for political power and voters, and academic disciplines compete with each other for research funds and commissions that will attract increasing numbers of able students. Thus every discipline needs a 'party programme', a vernacular definition, which functions outwards as the discipline's uniting programme.

We also have to realize there are basic differences between geography as a 'school discipline', as a 'university study' and as a 'research discipline'. The **school discipline** will not only be defined by academic geographers but, to a large extent, also by ministries of education, general educationalists and school politicians. The themes of knowledge and the school hours allocated to geography are dependent upon the roles of other disciplines and the public's expectations of the contributions geography is conceived as making to general education. **University study** at BA level is structured by academic university teachers, and principally it has two aims: to give prospective schoolteachers a general geographical education that must somehow cover the discipline's field as taught in schools; and, secondly, to form the basis for further studies and research.

Today it is a general requirement that university teaching should be research based. It is thus rather tempting for a professor to focus lectures around his or her recent specialist research activities, forgetting that most of the students will need jobs as generalists in schools or public planning. Lecturing based on personal research experiences might be very inspiring to the students but, as a consequence, important themes that are not currently the focus of research at a particular university may be left out. This is a basic conflict in any university discipline, but it may be particularly acute in geography because of geography's broad scope of themes. It is here that university teachers need to consider the discipline's vernacular side, without compromising academic standards and requirements. The term 'research based' also includes research done by others outside one particular limited specialism: university study cannot be structured solely on any one specific current departmental research activity.

The **research discipline** (in which departmental staff, PhD and, to some extent, MA students take part) has to be constantly changing and developing in line with research problems currently regarded as most promising by academic researchers. There should, in principle, be complete freedom to take up any kind of research theme, but this could create problems for the discipline if no regard is taken as to the history, traditions and public expectations of the subject.

Box 4.5   The popular notion of geography

The scene was typical of that extraordinary ritual known as the Cocktail Party. – Groping for something else to fill the silence, she got in her word first. 'And what do *you* do?' she said. 'Oh,' I said, grateful for the usual filler, 'I'm a geographer.' And even as I said it, I felt the safe ground turning into the familiar quagmire. She did not have to ask the next question, but she did anyway.

  – 'Oh really, a geographer . . . and what *do* geographers do?'

It has happened many times, and it seldom gets better. That awful feeling of desperate foolishness when you, a professional geographer, find yourself incapable of explaining simply and shortly to others what you really do. One could say, 'I look at the world from a spatial perspective, in a sense through spatial spectacles,' or 'Well, actually I'm a spatial analyst,' both of which would be true up to a point. But such phrases convey no meaning to most people, and leave them suspecting that you need a new oculist, or perhaps an analyst of a different sort. – In a desperate attempt to build a bridge with familiar words, one ends up saying, 'Well, actually, I teach geography.' 'Oh really?', and laughing. 'What's the capital of North Dakota?'

(from Gould, 1985, pp. 4–5)

This means that, on the one hand, we have to strengthen public understanding of our discipline, its developing research themes and knowledge base and, on the other, in some way adapt our academic research to public expectations. It may not be easy to strengthen public understanding if we have to start from scratch (see Box 4.5), which is the case in many countries where high-school geography is very rudimentary (for instance, in the USA and Norway). The situation is much better in the UK because of the traditionally strong position geography holds in the school system. In any case, to be able to create a link between the vernacular and the academic we must take the discipline's traditional focal points as starting points. This is even more true because

> these focal themes arguably underpin much subsequent debate, and in large part set many of these parameters within which contemporary research is conducted. These dual edifices, then, on which geography was built, are the human–environment relationship (or man–environment relationship to use the vernacular of the day), and the regional concept.
>
> (Cloke *et al.*, 1991, p. 4)

Geography has traditionally been the one discipline to bridge the division between social and natural sciences. It seems suicidal to discard that position today when there is a public need for both educational efforts and research to analyse natural and social factors affecting the relationship between humanity and the environment. It is this author's experience (as an author of high-school textbooks) that it is possible in schools to assess the effects of social and physical factors at one and the same time. In academic research it is, however, necessary to specialize to create something worth while. Based on an 'academic' definition, Asheim (1990b) maintains that it is Utopian and arrogant to believe that it is possible to integrate physical, human and social geographic research at an individual level. This might be true, but the humanity–environment relationship should be of crucial research interest to both human and physical geographers alike.

Johnston (1986c, p. 450) maintains that natural and social sciences cannot be integrated because they have different **epistemologies** – they are different forms of science. Bhaskar (1979) maintains, however, that transcendental realism permits a reconsideration of the possibility of naturalism – the thesis of an essential unity of method between the natural and social sciences. There certainly is a fair gap between certain positivistic natural science methods and postmodernist approaches in social sciences, but there is a range of choice in between.

Other reasons for disregarding the idea that there is an unsurmountable barrier between human geography and the geography of nature are as follows:

1) In agreement with Cloke *et al.* (1991, p. 2): 'We believe that much recent debate has rested upon unduly dogmatic assertions of points of difference, and that this has been to the detriment of many of the continuing threads of geographic endeavour.'

2) The assertion that human geography should be developed solely as a social science may well lead to its demise as an independent discipline: 'Human geography as an exclusively social science loses its distinctive identity – it competes with sociology, economics, anthropology – but on their ground, not on ours' (Stoddart, 1987/1996, p. 471). If it were to become a common misconception that geographers are only undertaking social research that could be carried out in any social science department, how could geography then claim the right to an independent status in university budgets? Both in the USA and some European countries there are examples of formerly prestigious geography departments that have been closed as a result of such factors.

3) For more than one hundred years, geographers have tried to build up both an academic and a vernacular definition of geography as a science that bridges the gap between the social and the natural sciences: 'Some parts of the world are virtually empty of people, others are desperately overcrowded and becoming more so. Geography is about precisely this: earth's diversity, its resources, man's survival on the planet' (*ibid.*).

   An 'authoritarian' definition of geography pinpoints this:

   a) Geography explores the relationship between the earth and its people through the study of *space*, *place* and *environment*. Geographers asks the questions of *where* and *what*, but also *how* and *why*.

   b) The study of *place* seeks to *describe and understand* not only the location of the physical and human features of the earth, but also the processes, systems, and interrelationships that create and influence those features.

   c) The study of *space* seeks to explore the relationships between places and patterns of activity arising from the use people make of the physical settings where they live and work.

   d) The study of the *environment* embraces both its physical and human dimensions. Thus it addresses the resources, sometimes scarce and fragile, that the earth provides and on which all life depends; the impact on those resources of human activity; and

the wider social, economic and cultural consequences of the interrelationship between the two. These three elements form the core of geography. Uniquely, they
create a bridge between the humanities and the physical sciences.

(National Curriculum Geography Working Group, 1989)

This might be seen as a 'party programme' or a **vernacular** approach to a definition. As it is meant as a framework for the discipline in schools, some might also
discard it as irrelevant to academic research. However, the academic discourses
around the 28th International Geographical Congress in The Hague in The
Netherlands in 1996 resembled this programme to a great extent (Box 4.6). In the
next chapter we will analyse how the above-mentioned basic concepts in geography (space, place and environment) are being used in recent geographic research.
We will also discuss such concepts as region, territory, local identity and the
spatial aspects of social exclusion and 'otherness'. Our starting point will be the
concept of **relational space**.

Box 4.6   Land, sea and human effort

'Land, sea and human effort' was the main theme of the 28th International
Geographical Congress in The Hague, 4–10 August 1996. The theme seems to have
two connotations, one general and one more special. 'Land, sea and human effort' is
in general what geography is about. In The Netherlands, however, this theme has a
special meaning, depicting in some way the core of Dutch identity: the human effort
to build up land through centuries of struggle with the forces of the sea. The
opening focus of the conference was on this theme – The Netherlands as created
and formed in a tedious ever-lasting struggle with the forces of the sea and the
rivers. Geography is an important university discipline in The Netherlands; at
Utrecht University there is even a Faculty of Geographical Sciences, not merely a
department. This faculty has courses that cover geology, geophysics, cartography,
GIS and natural and cultural landscape studies, as well as cultural, economic and
social geography. Dutch geographers are practical people and have been very active
in the development of the Dutch planning discipline: *Planologie*. In planning their
country, the Dutch have always needed a thorough understanding of both the
forces of nature as well as of transforming human agency. Even though there is a
strong tradition of social geography in The Netherlands, the humanity–land–sea
relationship cannot be forgotten as it is so strikingly present in everyday life and
survival.

   The author had the privilege of editing a special issue of *GeoJournal* for the 28th IGU
congress. This issue also took the main theme of the congress as its leading thread.
Geographically, the contributions covered the North Sea and the northern Atlantic
including their coastal areas, and the thematic focus was on coastal zone management,
management of ocean resources and the pollution of the seas. The intention was to
present geography's integrative possibilities by inviting contributions from both physical
and social scientists and from ecogeographers.

Some general lines of thought could be traced through the articles published in the journal:

1) *General geographic structures activate a set of mechanisms in local contexts which create specific localities* (see Figure 4.6, p. 131). Localities are not singular and we should not fall into the **singularity trap**; neither should we fall into the **generality trap**, since general structures are always filtered through local evaluations and conditions. When it comes to considering physical structures, this is rather over obvious: the sea as a general geomorphologic agent works differently on a debris coast (as in The Netherlands and Denmark) than it does on a rocky coast as in Norway. This makes human efforts to protect coasts necessary in The Netherlands and Denmark but almost superfluous in Norway (see Figure 4.10).

   On the human side structures are adapted to local conditions in other ways. In the fishery sector modernization is a general structural process, but it has effectively been modified in many fishing communities along the North Atlantic shores. Coull (1996), for instance, demonstrates how the Shetlanders have tried to counteract general trends by exerting stronger control over local natural resources in an effort to develop a sustainable island economy. In analysing coastal zone management, Bennett (1996) found that the local cultural context (particularly common codes of communications and traditions of co-operation) is of crucial importance for the success of local coastal planning. Bjørlykke (1996) shows how the structures of the international oil business have had to adapt to different local cultures, interests and administrative structures in the Shetlands, UK, and Øygarden, Norway. Lundberg and Handegård (1996) found when analysing local farming practices in the Norwegian coastal zone that even at the individual level of the single farmer there has been relative autonomy in relation to socioeconomic structural forces. Spatial activity is governed by individual mental schemata, which are based on social norms and relations, knowledge and experience of resource use, and adaptations to (or protests against) general economic and political forces.

2) *The local processes of change from vertical to horizontal linkages in resource use and dependency: the 'change of modernity' or 'threat of modernity'.* Lundberg and Handegård (1996) demonstrate how farming practice, land use and vegetation have changed in Norwegian coastal landscapes over 250 years, having adapted to shifting dependencies on local resources. Most coastal settlements grew up to harvest local fishery and land resources – the vertical (local) connections. **Modernization** has led to specialization as either farmers or fishermen. In Denmark, as discussed by Meesenburg (1996), larger boats that fish in more distant waters have led to a concentration of fishing activity in a few deep harbours like Thyborøn (Figure 4.10). The settlements that previously depended on open-shore fishing are now given over to tourists. The depletion of fishery resources has meant that coastal communities have had to find new sources of income to survive. Tourism has been an obvious solution for Danish and other attractive, former fishery-based settlements around the North Sea and the northern Atlantic. In both the Shetland Islands and in Norwegian fishery communes, fish farming has become a means of preserving local settlements. Salmon from the Shetland and Norwegian fish farms finds markets in Paris as well as in Tokyo. Horizontal dependencies are much stronger now than when the settlements were established, but the local environment and local resources are still important for both tourism and fish farming.

Figure 4.10 Air photo of the entrance of the Limfjord at Thyborøn in Denmark, with town, harbour, dykes and groynes. The photo demonstrates the human efforts to control natural processes; strengthening of dykes, dredging of the ship channel is continuously going on to keep the land and the fishing town intact. (Photo: Svend Toūgaard in Meesenburg 1996)

3) *The feeble balance between humanity's actions and their effects on natural processes and sustainability.* Overharvesting of fishery resources in the North Sea and the northern Atlantic and discussion of different measures and methods to regulate harvesting were naturally a core theme of the contributions to this special issue. It is a theme to which biogeographers, physical geographers, social geographers and economic geographers can all contribute. We need to understand the social structure of fishing communities as well as the economic and social effects of such measures as 'tradable fish quotas'. And we need to understand the changing oceanographic and biological factors that influence the developments of fish populations.

5

Processes in Space and Place

Of course the first thing to do was to make a grand survey of the country she was going to travel through. 'It's something very like learning geography,' thought Alice, as she stood on tiptoe in hopes of being able to see a little further.

(Lewis Carroll (1872) *Through the Looking-Glass and What Alice found There*

Space is not an empirical but a formal and classificatory concept. It is a frame of reference for the physical components of actions and a grammalogue for problems and possibilities related to the performance of action in the physical world.

(Werlen, 1993)

A whole history remains to be written of *spaces* – which would at the same time be the history of powers – from the great strategies of geopolitics to the little tactics of the habitat.

(Foucault, 1980)

Returning to memories of growing up in the social circumstances created by racial apartheid, to all *black spaces* on the edge of town, I reinhabit a location where black folks associated whiteness with the terrible, the terrifying, the terrorizing. White people were regarded as terrorists, especially those who dared to enter the segregated space of blackness.

(bell hooks, 1992)

The twentieth century has ushered in the discovery of *deep space* or at least its social construction, and yet it is only as the century draws to a close that this fundamental discovery is becoming apparent. Deep space is quintessentially social space; it is physical extent infused with social intent.

(Neil Smith, 1990)

For geography matters. The fact that processes take place over *space*, the facts of distance or closeness, of geographical variation between areas, of the individual character and meaning of specific *places* and *regions* – all these are essential to the operation of social processes themselves. Just as there are no purely spatial processes, neither are there any non-spatial processes. Nothing much happens, bar angels dancing, on the head of a pin.

(Massey, 1984, p. 52)
emphasis added

GEOGRAPHY AT THE TURN OF THE MILLENNIUM: ACTION AND PROCESSES IN SPACE

Since Alice started on her journey into the unknown land more than a hundred years ago, the discipline of geography has travelled a long way. A hundred years ago the geographer's job was the 'grand survey' – mapping the landscape, the

absolute space, as seen from a balloon or a hilltop – to guide subsequent travellers through it. On the basis of this surveillance, curiosity would lead on naturally to the basic geographical question: '*Why* is it like that here?' Alice did not pose that question; she took in the 'strangeness' of life on the other side as a matter of fact. Geographers, on the other hand, have had problems with this question since ancient times, having had problems in finding appropriate methods to analyse the why of the things observed. The French school of regional geography found a solution in **temporal analyses**: a historical explanation of the single instance, of how the present regional landscape and culture have been moulded in the long interplay between humanity and nature. This seemed an appropriate method in societies that had a local resource base, in which **vertical linkages** predominated. In this approach space remained, in Werlen's (1993) terminology, a 'classificatory concept', a 'container' more or less closed off from the outer world, with characteristics that had to be explained by the relations and processes within it. Each region ('container') is considered different from the others, and it is the geographer's task to explain the local basis for these differences. This approach is based on **total-special reasoning**, which Schmithüsen (1976) maintains is peculiar to geography. Its aim is to understand the integrated complexities that coexist in a part of the earth. It would be better to call this approach **chorologic** – in many regional projects, analyses of processes and relations between factors give convincing explanations for the characteristics of the single region.

Globalization (or internationalization) has, however, turned most societies 'upside down': **horizontal linkages** seem dominant in most societies today. This forms a basis for the **spatial science** aspect of geography, the search for spatial structures or models that, through simplification, could be idealized as geometrical models (see Figure 3.13, p. 84). **Relative space** (the location of, and distance between, different phenomena in the structure) became the focus of investigation. Distance as measured in terms of transport costs, travel time, mileage through a network and even as perceived distance has been accorded explanatory power. Geography's aim, therefore, was to formulate spatial laws. However, Werlen (1993) maintains that the notion of relative space was a blind alley, as it suggests that space as such has explanatory power. Space is to his mind only a frame of reference for processes and actions; distance as such cannot explain human actions. Human **interpretations** of distance and other factors in the material world are a better starting point.

The methods of the **spatial science school** have been labelled 'spatial fetishism': spatial correlations were investigated in a traditional positivistic way, just as empirical generalizations look for regularities at the level of events (see Figure 4.6, p. 131), isolated from the structures and processes of which they are a part. The preceding chapters have discussed reactions to geography as a spatial science, including those schools of thought that sought explanations of spatial structures in such non-spatial social sciences as psychology, economy and sociology. Discourse has centred on whether spatial structures can only be interpreted and explained via the single human being (**individualism**) or as an interpretation of the structural forces in society (**collectivism**). Basically, it is possible to talk of two traditions: one concerned with human intentions and meaning and the other with structuralist and Marxist approaches, focusing on systems and structures. Fosso

(1997, p. 23) points out that both individualistic and collectivistic methodologies have some merit, as society is both collective and individualistic: on the one hand, we live in a structured world with limited scope for different choices of action; on the other, we live in a privatized and self-structured world with room for self-reflection and conscious choice. This is also Giddens' message. Spatial structures, however, recede out of focus; space is seen either as a passive reflexion of the structural forces in society or as a conceptual framework in the individual's mind (phase 2 in Figure 5.1).

As space cannot be considered a separate factor in explanation, Werlen (1993) maintains that **actions** rather than space should be the focus of human geographic inquiry. Actions are, however, actions in space. Spatial differences, whether material or immaterial, influence social and natural processes. Werlen (*ibid.*, p. 79) refers to the three worlds in the process of action, as presented by Schutz (1982):

- *The subjective world* (World 1), consisting of the agent's stock of knowledge and which is formed from specific biographical elements.
- *The social world* (World 2) into which the agent is socialized and which is constituted and reconstituted as products of actions. This world also contains the 'objective' meaning-contents and consequences of the actions of others.
- *The physical world* (World 3) which consists of the material facts of the outer world, including the agent's body.

The physical world and the social world both exist independently of ego. But

> Schutz's realism postulate is a relative one. All reality exists for the knowing self only in the form of its meaning as it appears to the subject, and not as it might exist as such. Husserl and Schutz do not deny the existence of 'objective' facts, but place them in a different context. They do not assign them the status of an existence independent of the agent, but see them as the product of the subject's acts of cognition, as facts whose meanings are intersubjectively constituted. In other words 'objective meaning' signifies for Schutz something akin to the 'generally valid meaning' of fact or circumstance, to an intersubjectively shared meaning-context.
>
> (Werlen, 1993, pp. 93–4)

The human agent responds to stimuli (selectively received from the social and physical environment) and, *vice versa*, his or her intended and unintended actions influence the physical and social worlds (see Figure 4.7, p. 133). Werlen (*ibid.*, p. 10) points out that the **behaviourist approach** to social geography basically makes 'space' into a cause: the focus is on individual 'mental maps' and 'spatial awareness', whereas from an action perspective things would look differently. Although many actions are routinized and so not consciously planned, the main focus of action-orientated research would be on reflexive and intentional activity. Behaviour theory explains human activity as determined by stimuli, whereas according to action theory human activity is intentional and meaningful.

RELATIONAL SPACE

The most crucial question is: how we can argue that spatial differences are important but that space itself is only a formal and classificatory concept? Fosso (1997) maintains that the answer is to be found in the concept of **relational space**. Basically this means that space is contained *in* objects in the sense that an object

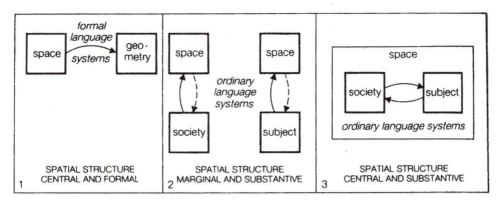

Figure 5.1   The changing roles of spatial structures in geography, phase 1, 2 and 3
Source: Gregory in Johnston, edit, 1986

can be said to exist only as far as it contains and represents within itself relation-ships with other objects. A site has a value, not because of its own isolated existence but because it contains relationships with other sites (Harvey, 1973, p. 14). In contrast to absolute and relative space, relational space can thus be re-garded as a kind of object characteristic. A better understanding of this may be gained if we link this concept to **action theory**: 'If "action" rather than "space" becomes the central theoretical concept of geography, the spatial *arrangement* of objects becomes relevant as a necessary condition and consequence of human action rather than a cause – the agents themselves are using different frames of reference for their orientation in the physical, subjective and social worlds' (Werlen, 1993, p. 143).

In an objective sense, agents orientate themselves both in the physical and social world by means of 'mental maps'. On these mental maps agents mark out the positions and paths by which they can achieve their goals. The physical world forms fixed points (for instance, 'home' and 'workplace') for social interactions, whereas the social world forms a spatial frame of reference through dimensions such as 'religion', 'science', 'language', 'art' and 'ethics'. This spatial frame of reference is contained in the agents and their actions. Space and time are the basis for all existence. In discussing the educational choices and migratory patterns undertaken by young people, Fosso (1997) suggests that the crucial question is how space in different ways is embedded in and through the actions of the young people, and how the products of such actions produce and reproduce different spaces. This embedded space is **relational space** – space is a social phenomenon, in as much as all social phenomena have spatial extention, localization and meaning.

It is thus possible to speak of the uniqueness of place, not as an ontological fact but as an epistomological contextual reality that is found in human actions and everyday life. People live their lives locally, but their consciousness is formed in and by different spatial contexts. Figure 5.1 presents phase 3 as an illustration of relational space, in which spatial structures are directly embedded in both social structures and individual actions. Phase 2 represents the period in which structuralism   and   humanism   relegated   space   to   something   secondary;

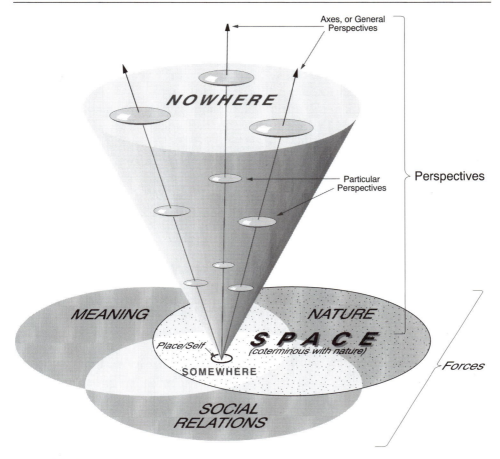

Figure 5.2 The relational geographical framework
Source: Sack, 1992, improved by Sack, 1999

and phase 1 represents the spatial science school of thought. The phase dominated by the notion of absolute space or regional geography is not included. It should be noted that, as all models, this is a simplification: the best studies in traditional regional geography involved an understanding of (although the term was not used) relational space in their studies of local ways of life, i.e. within the 'container' of absolute space:

> I regard as a fundamental theme . . . that social theory *must acknowledge, as it has not done previously, time–space interactions as essentially involved in all social existence* . . . Social activity is always constituted in three intersecting moments of differences: *temporally, paradigmatically* . . . and *spatially*. All social practices are *situated* activities in each of these senses.
>
> (Giddens, 1979, p. 54, emphasis added)

Giddens backs the concept of relational space, but geographers like Paasi (1991, p. 243) criticize Giddens' use of the spatial concept '**locale**' as it refers only to the use of space as providing the physical setting for interaction. What is lacking is the historical constitution of locales as this is formed in an intertwined process between the social and the spatial. Space and time have basic roles in the constitution of **everyday life** and, through the medium of human actions, spatial structures are produced and reproduced.

Sack (1992) has developed this relational concept of space, and its main features are given in Figure 5.2. Here space and place constitute and integrate forces from the **realms of nature**, **meaning** and **social relations**. Sack points out that, in addition to these three, it is possible to talk of a fourth realm, **the realm of agency**. This is based on the claim that the forces from the other realms are not determinate; we have power to choose for ourselves. This realm, however, is not included in the figure as it cannot be seen separately from the other three.

Box 5.1   Existentialism, Heidegger and 'being' in the world

Basically, many scientists see existentialism as 'anti-intellectual philosophy on the grounds that one of its central tenets holds that reality and existence can only be experienced through living and cannot therefore be made the object of thought' (Walmsley and Lewis, 1993, p. 117). On the other hand, existentialism – particularly through the philosophy of Martin Heidegger (1889–1976) – is crucial for an understanding of relational space as presented by Sack (1992). The central concept in Heidegger's philosophy is *Dasein* (i.e. 'being here'), a German word meaning a human subject's existential 'being' in the world. As human beings we are, on the one hand, interwoven with the environment and processes that make up the world and, on the other, we 'step out of' this unity to observe, experience, reflect on and choose between possible ways of being in the world. This duality means that every one is, to varying degrees, estranged, alienated from the world. A schizophrenic seems to live in a deep existential fear that alienates him or her from a meaningful being in the world. We are all located on a scale between 'being' and 'not-being' in the world. The human struggle against estrangement is essentially connected to our existence in space/place (being) and time (becoming). Every human tries to eliminate detachment through the creation of meaning in places of 'being' in the world and through a meaningful future life of 'becoming'.

The world is a structure of meaningful relations in which the individual exists and which he or she partly creates. Heidegger distinguishes between three aspects of the world: the physical, the social and the subjective. We can link these three aspects to the realms of nature, social relations and meaning in Sack's figure (Figure 5.2). As human beings we are 'thrown' into the physical world or realm of nature with the conditions set by the place we live in and the material conditions that set frames for our future. An important part of 'being' in the world relates to how we manage to cope with these conditions.

Another crucial part of being is our relations to, and dependencies on, other human beings. Basically, however, we are all alone: no one else can live my life or die my death, so to be in the world we need a subjective meaning for our existence.

As place and space are intrinsic parts of our being in the world, our individual relational space is the basis for our actions. If we are to understand how this relational space influences humanity's being and becoming in the world, we need to encounter people and situations in an open, intersubjective manner, which demands we use *qualitative methods*.

There are a number of theories about how the forces from one or other of the realms can influence us (Box 5.1). The **realm of nature** consists of the elements and forces described by the natural sciences, including physical geography. As humans are part of the natural world, it is a basic assumption that we are governed by this realm's forces. While agreeing to some extent with Werlen (1993) that it is difficult to attribute an explanatory force to space in the realms of social relations and meaning, Sack maintains that this is not the case in the realm of nature. Sack (1992, pp. 80–1) points out that spatial variables, such as distances and lengths, are prominent in all areas of natural science. But it is still difficult to regard space as a causal agent: the concept of relative space has not brought us closer to explanation, and absolute space still seems relevant. Sack suggests that the realm of nature should be described in a modified relativistic sense, or **relationally**. Relationally here means that the properties of space are absolute and describable in Euclidean terms. Space exerts effects but not absolutely because, on earth, space is not empty, and its effects are mediated by matter or substance. For instance, the time it takes to travel is dependent on the mode of travel as well as on the physical distance.

The **realm of social relations** analyses how such factors as class, government or modes of production influence us. In studies undertaken by social scientists, space is often not considered as part of social relations. Sack (*ibid.*, p. 82) points out, however, that by two related means space becomes an elemental part of this realm. By adopting the concept of **relational space** it enters the social realm, as social relations are about individuals and their connections with one another in space and time. This is the same concept of relationality as applied to the natural realm. The second way of introducing space into this realm is through the concept of **territoriality** – social influence can be achieved by controlling area.

The **realm of meaning** is linked to theories that seek to describe how our ideas, values and beliefs have particular forms and influence our actions. In his earlier work, Sack (1980) demonstrated how space and time help in constituting different types of meaning: 'For example, different artistic forms or meanings result from different combinations of space and time. Science, ritual, myth and magic and everyday life can be analyzed in terms of how they use spatial and temporal intuitions' (Sack, 1992, p. 82). It should be added that, basically, space is here a formal and classificatory concept but, as agents are acting and relating to the world around them, space is given relational and symbolic meaning. For instance, Sack (*ibid.*, p. 16) points out that ghettos and slums contain not only social relations but also physical structures and meanings. To most, these are undesirable places that, through human agency, may be perpetuated or transformed into something else. The **realm of agency** is the central focus of everyday life since it combines the impulses from the other three realms.

In intellectual life – in the scientific division of labour – the realm of nature, the social realm and the realm of meaning are, however, separated, somehow creating an intellectual deadlock. Here, maintains Sack (*ibid.*, p. 15), the concepts of space and place are essential categories that incorporate all the realms. The cone in Figure 5.2 (starting in 'somewhere' and extending out to 'nowhere') is intended to indicate two things:

1) 'Somewhere' (indicating 'insidedness') and 'nowhere' (indicating 'outsidedness') are limiting cases and are never in themselves completely attainable. 'Somewhere' is the personal perspective of the 'ego', whereas various degrees of public, abstract or objective 'outside' perspectives are located further up in the cone.

2) There are many paths from 'somewhere' to 'nowhere', even though they are inter-related. The lenses in the cone represent such different paths or perspectives. A religious, moral lens may, for instance, be located rather close to 'somewhere'. A scientific, abstract lens closer to 'nowhere' could draw attention to place as a location in space, whereas a less abstract lens further down on the same axis could be analysing a personal sense of place (*ibid.*, p. 18).

Throughout the history of geography, most approaches have been close to 'nowhere'. Sibley (1995, p. 185) points out that even now there is

> an often excessive distance between the observer and the observed which means that the categorial schemes of the former are unchallenged by the experience of the other and by other world-views. The voyeuristic impulse, which Gillian Rose represents as the 'masculine gaze', reifying and dehumanizing 'the other', has a particular expression in cartography and other forms of distant geographical description.

Sibley reminds us that mapping provides us with security but that the same voyeuristic danger is inherent in much analytic work on the texts of 'father-figures'. Debates on the texts of such 'father-figures' as Marx, Giddens, Foucault, etc., are intellectually safe, and they provide as well an élitist and hierarchical view of knowledge. Sibley asks geographers to go out into the world to experience the life-worlds of other people.

The concept of **relational space** is crucial if we are to obtain a better understanding of the geographies of exclusion, of gender relations, territoriality and 'otherness'. (In Box 5.2 we take a fresh look at Alice's experiences in Wonderland, which shows she did experience – in a rather tough way – the relational aspects of space and place.)

Box 5.2   Relational spaces in Wonderland

Alice started from 'somewhere' in her family garden in Oxford when she decided to follow the White Rabbit. But was she going 'nowhere'? A number of different perspectives are juxtaposed in her experiences in Wonderland:

'Will you tell me, please,' she said, 'which way I must go from here?'
  'Yes,' said the Cat, 'but mustn't you tell me where you want to go?'
  'Well, any place –' Alice began.
  'Then you can go any way,' the Cat said.
  '– if it is a place,' Alice said.

After pointing out the Hatter's house, the Cat explained that if every one in Wonderland were mad, Alice must be mad too. The Cat did not go away, but it was still not there anymore. It just disappeared, its big grinning mouth the last to go. Space and time do not matter in Wonderland. At the Hatter's house, the watch does not tell the time because it is always tea-time, and Alice is told there is no place for her although there are many empty chairs. And strange doors and paths appear at the most unlikely places. Luckily she has pieces from both sides of the mushroom to eat to get smaller or bigger when it pleases her. This comes in handy when she enters the rose garden to play croquet with the Queen of Hearts. This is the Queen's **territory**, over which she seems to have total control, shouting 'Off with his or her head' every time someone displeases her. But heads were not cut off, as it was difficult to find the head of a playing card or the body of the Cheshire Cat. The trial before the Queen's tribunal is fearsome as long as Alice is the size of a small girl; but as she swallows a 'get bigger' bit of mushroom the assembly is reduced to a pack of fifty-two small playing cards.

This strange children's book gives us some clues to the concept of relational space. All through the book competition over territory is demonstrated, as well as the fear and attraction of unknown spaces behind closed doors. Gradually Alice gains control over relational space and her fears disappear as she learns how to change her size. Her relational space, her relationship to the physical and social realms as well as the realm of meanings in Wonderland, is changed.

Whereas there might be disagreement as to the value of the 'spatial' in geographical explanation (and particularly the notion of geography as a '**spatial science**'), the concept of '**place**' may symbolize the essential aspects of being in the world. As noted by Sack (1992, p. 80), the sense of personal place 'that is rich and complex and possesses something of a holistic quality seems to take primacy over the sense of space: from the personal perspective, space is little more than a sequence of places'. This is a good reason to look at geography as the study of **places**. However, our starting point is a retrospective look at the **regional concept**.

REGION AND PLACE

Palm (1986, p. 473) presents 'the basic strength of geography as a field that seeks synthesis of human action and environmental structure in order to understand the complexity of place'. Thus geography is about what places are like, or about the variety on the earth's surface. Throughout the twentieth century the idea of **regional geography** has been connected with the classic presentation of geography as a discipline of synthesis (Holt-Jensen, 1988, pp. 125–6). A clear-cut defence of regional geography was given by John Fraser Hart (1982) in his Presidential Address to the Association of American Geographers in 1981 under the title of 'The highest form of the geographer's art'. Others, too, have supported some form of revitalization of regional geography. Aase (1970, p. 13) asks whether geographers will not be required to discuss the development of a new regional geography, employing new methods of attack. He points out that planning a modern society requires a good deal of research on the development of synthesis, but that currently this work is almost wholly in the hands of economists and architects – leaving geographers on the sidelines with their exclusive concern for **systematic** geography. Modern geography can, however, develop methods that can be used in the formation of regional synthesis. Taylor (1985) envisages a third (pure) geography, which takes much of its inspiration from traditional regional geography. But writing regional geography presents some major problems, says Johnston (1991, pp. 41–2, based on Paterson, 1974):

1) The problem of *verbal* presentation of the complex coexistence in space of disparate phenomena. We may take a photograph of a landscape from a particular viewpoint, but to explain in words even the physical features of the landscape is difficult. In 1910 E. M. Forster described a viewpoint (199 m a.s.l.) from the eastern part of the Purbeck Hills overlooking Bournemouth and the vales south of Salisbury Plain as a climb which is the 'wisest course to show a foreigner England'. Taking a fresh look at Forster's description, Bird (1993, pp. 60–3) suggests that a verbal account of the view will have the form of either 'reason' or 'geographical imagination'. Reason will attempt a factual account, which is rather dull and difficult to structure. Forster's novelistic description is imaginative and vivid. This is the *verstehen* type of presentation, which very few have the gift to undertake.

2) The problem of finding an approach that highlights the essentials of a region. Ideally, each region should be treated in its own special way but, very often, it is safer to use a standard format, which will rapidly lead to uninteresting accounts.

3) The difficulties in defining a region. Few regions have clearly defined boundaries. This is a particular problem with **uniform** and **natural** regions. In such regions the borders of a number of features are supposed more or less to overlap to create an area that is uniform within, but clearly distinctive from, its surroundings. In **functional regions**, the borderline is where the influence of one centre diffuses to be overtaken by the influence of another centre.

4) The problem of scale. The regional geographer must find a level of generalization that will neither drown the reader in detail nor leave him or her with the feeling that he or she still does not have a clear image of the region in question (Paterson, 1974, p. 11).

5) The growing shortage of local data. Although there has been a great expansion in geographical investigations, statistical records, etc., much of this is not available at the local level. Because of stricter rules, it is increasingly more difficult for researchers to obtain access to local primary data, and as the processes of change are very fast these days you cannot be content with material collected some years ago.

6) The submergence of regional distinctiveness. The **globalization** of the economy and the internationalization of all forms of human activity (as, for instance, architecture and transport systems) remove variations and distinctive features. A region can no longer be explained as the result of **vertical** humanity–land relations; **horizontal relations** will link any region to large parts of the world.

Paterson (1974, p. 21) believes that these six problems can be resolved, but his argument is defensive, particularly in suggesting that the goals of regional geography 'are general rather than specific . . . popular and educational rather than practical or narrowly professional'.

Johnston (1991, p. 42) advocates another approach – not for a separate activity in the form of regional geography but for a 'geography that is regionally informed'. This is based on what was in the 1980s termed **'new regional geography'**: the 'recognition that spatial variations are fundamental to the organisation of society, with the world comprising a complex mosaic of specific places within which general processes are enacted, but whose features cannot be accounted for by those processes alone' (*ibid.*, p. 49).

Gilbert (1988) suggests that 'new regional geography' is concerned with regions as

- local responses to capitalist processes, which was particularly prominent in the UK in the 'localities research programme' (see Massey, 1984; Cooke edit, 1989);
- a focus of identification (or 'sense of place'), which was a particular focus for French analysts of cultural identities; and
- a medium for social interaction, which also involved analyses of geographies of exclusion (see Sibley, 1995, and pp. 171–75).

These concerns are rooted in both **structuration theory** and **realism.** Regions are not singular; they are influenced by structural forces but the outcomes are **contingent**, influenced by local situation and the choices of human agents. The new regional geography focused on 'the persistence of regional diversity in the face of the homogenizing tendencies within capitalism [which] provides regional geography with practical significance' (Johnston, 1997, p. 249).

Massey (in Massey and Allen, 1984) had already maintained that 'geography matters' when she advocated the 'new regional geography' and hence pointed to the role of the local milieu in structuring people's lives and decisions. A number of terms have been used for this milieu: place, region, locality and locale (the last being a reference to Giddens (see p. 127)). Johnston (1991) prefers to use the term **place**, which we will adhere to here: 'Thus cultures develop in places and are passed on in places. They are continually changing and have no existence independent of the people who use them' (*ibid.*, p. 50). We are all socialized into places and are, in spite of modern communications, bound to home and work in a space-time prism (see pp. 81–82), and we are rooted as social beings in places we have grown up in and have developed our social life. Our actions may be understood partly as reflections of our social networks (or relations to others) in bounded places, and partly as reflections of our own development and roots in places that have formed us. Massey and McDowell (1994) give an effective demonstration of this in their essay 'A woman's place' (Box 5.3).

Box 5.3   A woman's place?

Massey and McDowell (1994, pp. 191–210) point out that the growth of capitalist forms of production in the nineteenth century created a spatial division of labour in Britain, in which different regions played different roles. Capitalist relations in production were accompanied by other changes, particularly changes that disrupted the traditional relations between women and men. Capitalism presented patriarchy (male dominance) with different challenges in different parts of the country. These changing relations were studied in four areas:

- The coal-mining areas of northeast England.
- The factory working areas of the Lancashire cotton towns.
- The sweated, homework-based sewing areas in inner London.
- The agricultural gangwork areas of the fenlands.

Life in colliery villages could be described as being dominated by male solidarity and female oppression. The male was the bread-winner and did dangerous work, which created strong solidarity between workers. Women had few possibilities for paid work but had a hard time serving the men, who worked shifts and demanded hot meals at all times of the day and night. Dirty clothes had to be washed and the miners' backs scrubbed before the men could assemble at meetings or at the pub. This system of male dominance remained unchanged until the middle of the twentieth century.

On the other hand, factory life in cotton towns created many opportunities for women to obtain paid work outside the home. Spinning and weaving factories were a direct challenge to the traditional sexual division of labour, which made it more difficult for men to maintain their patriarchal domination. Spinning, however, had traditionally been women's domestic work but, in the factories, it was classified as 'heavy' work and taken over by the men. In the weaving industry, on the other hand, women and children were mainly employed. As women received their own wages their self-confidence grew. As Engels (1969, p. 173) pointed out, in many cases family relations were turned 'upside down. The wife supports the family, the husband sits at home, tends the children, sweeps the rooms and cooks.' Because of this, Lancashire women workers joined trade unions more frequently than women in other parts of the country, which also created highly localized support for the middle-class-based suffragist movement.

The rag-trade industry in Hackney, inner London, also created jobs for women, but the workplace was the home. Sewing was something that was traditionally regarded as women's work. From a man's point of view, homework had its advantages: the women-folk stayed at home and the sewing jobs had to be completed in between other domestic duties. Hence in the sweated trades of London, capitalism and patriarchy conjoined to produce a system that was less of a threat to male dominance.

In the large fenland farms of East Anglia, there was a need for a workforce that consisted of the rather large, landless, agricultural proletariat. Women were an integral part of this workforce, doing all kinds of heavy work. In nineteenth-century England this provoked moral indignation, as did the cotton-mill employment of women: it was often necessary for the workforce to travel in working gangs from farm to farm, which engendered an independent attitude of mind that was condemned by the Victorian establishment. Women's waged labour seemed, however, not to present any threat to male supremacy here. One reason was the seasonality of the work; another that men, too, were predominantly agricultural workers: 'Women served their menfolk, and both men and women served the local landowner, nobody rocked the boat politically' (Massey and McDowell, 1994, p. 201).

It might be thought that these local differences would disappear with the advent of a postindustrial society: all parts of the country are now linked by mass communications (TV, radio and the national press) which imprint a national culture that is believed to wipe out regional differences. Massey and McDowell maintain, however, that regional differences remain, although not as pronounced and in a different manner. A great deal has changed since the nineteenth century, since over 40% of the national paid labour-force in the UK is now female. However, the gender division of labour has changed in different ways in different areas. One reason for this seems to be the particular social and economic history of women in each of the places Massey and McDowell studied.

In the coal-mining districts of the north east, more pits have closed down while new jobs, particularly for women, have been created in service industries and light manufacturing. It seems such industries are attracted to this region as women here have no tradition of waged labour and little union experience. The jobs are classified as semi-skilled and unskilled and so are low paid. Male outrage over the new gender division of labour has been vociferous, with male pleas for 'proper jobs'. The dominant male working-class culture is still evident in pubs and clubs, but even here male exclusivity is threatened. It might also be noted here that the same type of story is told humorously for the 'steel town' of Sheffield in the film *The Full Monty*. A career as a male stripper is both a desperate solution and symbol of new gender relations!

In the fenlands, there seems to be continuity rather than change in gender relations: agricultural work still dominates the area and attitudes to domestic responsibilities remain traditional. In the Lancashire cotton towns, women's employment has been declining for decades. With its pool of skilled, female industrial workers, Lancashire would seem a good place to establish new light industries and yet the new industries of the 1960s and 1970s did not come here to the extent they went to other places. The explanation is complex but seems to combine patriarchal and capitalist structures. The cotton towns were never given development-area status and the economic assistance connected with this, although reductions in employment were as large as in the coal-mining districts that received such a status. The reason could be that women's unemployment is not considered as serious as male unemployment, and a self-confident, independent female workforce is similarly unlikely to attract new investment. Finally, in Hackney, London, the home-sewing trade is still pursued but now in capitalist competition with cheap products from the Third World. The women of Hackney have become in part victims of the changing international division of labour and have to produce more for less to remain in business. If suitably educated, they may have the possibility of taking other jobs, but most are still caught in traditional gender traps and work conditions.

These examples show that societies change but that local conditions continue to be of importance.

Source: Based on Massey and McDowell (1994).

In their book *The Power of Place* (1989), Agnew and Duncan argue that place has been 'devalued' in 'orthodox social science' in recent decades as research has focused on the modernization processes that are wiping out regional differences leading to the development of a 'placeless society' – a process seen as universal, natural and lawful. The goal of *The Power of Place* was to 'argue against the prevalent tendency in history and social science to overvalue the social imagination at the expense of the geographical' (*ibid.*, p. 7). Their argument is supported by a number of case studies. These demonstrated that

- the perspective of place makes it possible to analyse such abstract categories as 'class' in an everyday context. It was found that 'class' (just as **'gender'** and religious affiliations) means different things in different places;
- based on a **realist** approach, it is possible to focus on the uniqueness of places and still be committed to causation. The structuration of social relations in **everyday life** contains many similar elements in different places (e.g. class, gender relations, central–local government relations, etc.), but outcomes differ from place to place; (see box 5.3)
- a focus on place recognizes human actions as motivated and intended but, at the same time, both are mediated by social structure and generative of it;
- a recognition of difference between places problematicizes models that divide history into phases. Societies develop differently and at different rates in separate localities, although influenced by the same structural forces; and
- local differences reflect cultural variations that are not merely **superstructural** reflections of the **'economic base'**. Rather, it is the 'practical nature of everyday life' (Agnew, 1987, p. 43) which creates the context within which people act, and which is the environment they re-create through their actions.

What then is a **place**? Following Aristotle, a place could be defined as a portion of geographical space occupied by a person or thing. According to this definition, a place refers to an exact restricted part of space defined by a particular content. For instance, the place of a book is the space this book occupies; the place of a town is the area containing its buildings, roads, etc. (Relph, 1996, p. 907). Such an interpretation implies, however, that place can only have physical, spatial attributes, and no particular content. Such an understanding of place could be useful in a spatial science context: places would form nodes in a spatial network (see Figure 3.13, p. 84). But such an understanding of place is rather mechanical and shallow. Place should rather be defined as the 'location of experience, the container of shapes, powers, feelings and meanings' (Walter, 1988, p. 215). This definition of places as **territories of meanings** rather than as containers of things takes its inspiration from **phenomenology**. As Figure 5.2 has indicated, however, place unites the realms of meanings, nature and social relations.

A SENSE OF PLACE

The Norwegian architect Norberg-Schultz (1984) understands **place** as the physical setting in which local social life takes place. A place is, however, not only a part of geographical space in which something is located. People live in a place and in the meeting between these people and the physical environment is created the spirit of place, or **'genius loci'**. Norberg-Schultz argues that 'place is evidently an integral part of existence', which can best be understood phenomenologically. For Nordberg-Schulz, it is basically the physical structure that defines 'genius loci'. Through an examination of such places as Prague, Khartoum and Rome, he asserts that these cities have strong spirits of place because they have strong visual properties and, as such, have a sense of mystery.

In the past, settlements were always clearly defined, closed units (compact village, medieval walled city); it is this figure or physical quality that gives meaning to the word **place**. Modern architecture and town planning have created physical structures that, in most cases, lack distinctive character. Modern towns are often monotonous and lack 'genius loci'. It is the architect's job to employ a methodology for 'place analyses' that will reveal the 'natural physical features' which are characteristic of special landscapes and historic settlement patterns. Physical planning should be based on these 'natural physical features' so that it creates places with definite identities.

In a similar way, the Swiss geographer Gutersohn (1945) advocated that planning should be based on analyses of 'harmony in the landscape', and he presented visual examples of harmonic and disharmonic settlements. What creates the identity of place for the people living there is, therefore, a learning process: people learn what constitutes a harmonious settlement and building pattern – social learning about what is nice or ugly. Each settlement has a distinct location in, and in relation to, a physical landscape which must be understood and respected. Further, most settlements have a history that underpins the 'genius loci' as understood by the inhabitants. New additions to the settlement must adapt to that historical connection but not 'freeze' it. Thirdly, every settlement needs an identity that transgresses distinctive character and historic connection: it needs both to appear as a distinct unit in the landscape and to have an understandable inner structure:

> The term *'sense of place'* is often used to mean the same as *'spirit of place'* [genius loci]. This is confusing. It is more appropriate to understand 'sense of place' as the *awareness* of spirit of place, and as a faculty which individuals possess rather than a property of the environment.
>
> (Relph, 1996, p. 910, emphasis added)

Eyles (1985) identified four different senses of place based on his analysis of residents' attitudes to their small English town. One of these senses he labelled 'apathetic' since some people had little interest in their surroundings. A second he called 'social sense of place', which is based on social relations: a place is identified through people's social relations with family and friends. The third is called the 'instrumental' type, which regards place primarily as a provider of goods, work and home. The fourth is the 'nostalgic', which stresses heritage (including old buildings). This analysis suggests that the sense and spirit of place are not related simply to the physical environment. Shurmer-Smith and Hannam (1994) see places as social constructions: something that is created intersubjectively between humans and not an objective reality. Even if we do not hold extreme opinions we need to focus more on the social sense of place than architects such as Norberg-Schulz do. We may define a place as the location of experience: the container of shapes, powers, feelings and meanings. Entrikin (1991, p. 6) points out that the **geographical concept of place** 'refers to the areal context of events, objects and actions. It is a context that includes natural elements and human constructions, both material and ideal.' Architects and natural scientists tend to regard place only as physical setting, whereas social scientists see place as an arena for social relations. In geography there are different views of place, which partly reflect a particular specialism as, for instance, social, physical or ecological geography. In broad geographical terms we will, however, maintain that the **identity of place** contains three essential elements: *physical setting*, *human activity* and *meaning* (or, in the words of Sack (1992), realms of meaning, nature and social relations) whereas **sense of place** may be seen as the individual awareness of such an identity.

Tuan (1996, p. 447) defines a place as a **node of activities**:

A street corner is a place, but the street itself is not . . . a street is directed, historical space on the horizontal plane, only non-directed homogeneous spaces can be places. When a street is transformed into a centre of festivities, with people milling around in no particular direction, it becomes a non-directed space – and a place.

There are considerable variations in both the type and intensity of **sense of place**, points out Relph (1996, p. 910), depending on such things as familiarity, detachment, social status, gender and self-consciousness. Eyles (1985) suggests that women connect place more with social community than men, who relate place more to its physical setting. Interviews with youngsters in an urban area in Bergen, Norway, revealed that the girls were primarily concerned with social relations and fear of violence, whereas the boys were more concerned with physical possibilities for soccer and other sports. Sense of place could thus, to a large extent, be **gender** related.

As Relph (1996) suggests, sense of place depends very much on familiarity: there are obvious differences between an insider's and an outsider's experiences. **Insidedness** is connected to often unself-conscious knowing and being known somewhere; **outsidedness** is a relatively detached relation to a place as, for instance, based on factual information a tourist may receive on a short visit. There are, of course, varying degrees of both insidedness and outsidedness, which means there is a gradual transition between the two. Developing a receptive and self-conscious sense of place is a particularly essential skill for geographers.

Terms such as 'identity', 'rootedness', 'at homeness' and 'symbol of self' have been used to describe the human sense of belonging to a place. Geographical research has tried to grapple with the individual sentiment and symbolism associated with place: both positive and negative psychological bonds with, for instance, the home environment have been explored. '**Home**' is mostly a place of security and comfort, a place in which an individual develops a conception of self (Walmsley and Lewis, 1993, p. 226): 'My home is my castle, but who is *me*?' Homes are occupied by families in which, sometimes, there is a bullying tyrant who makes it impossible for the other family members to develop an individual conception of self. Gendered stories often tell of homes as places of fear. When we start to analyse 'geographies of exclusion' and the life-worlds of those who do not manage to cope in the housing market, we encounter different feelings of 'home'.

Home is only one type of place we relate to; others are places we visit frequently, such as the workplace, a shopping mall, a school or a more distant summer-vacation place. Places also have different scales: your dwelling and the neighbourhood in which your home is located are places; you may also describe a town as a place. Although it is difficult to delimit clearly an appropriate scale for the concept of place, place is definitely at the lower end of the scale in the hierarchy of social spaces in our **life-world** (see Figure 5.3). A larger area might be termed a **region**: an urban region or the Pennine region.

The terms **topophilia** and **topophobia** have been used by Yi-Fu Tuan (1974) to describe human emotions towards place. Topophilia describes human love for a place. This can be the gentle and long-lasting love for a place called 'home', or the place where you grew up and have your roots. Alternatively, such a love might be created by a once-in-a-lifetime experience – such as this author had, experiencing

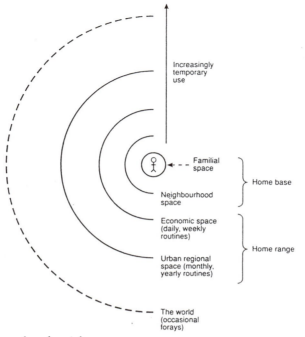

Increasingly
temporary
use

Familial
space

Home base

Neighbourhood
space

Economic space
(daily, weekly
routines)

Home range

Urban regional
space (monthly,
yearly routines)

The world
(occasional
forays)

Figure 5.3   A hierarchy of social spaces
Source: Porteous, 1977, p. 92, Walmsley and Lewis, 1993, p. 225

a sunset with hundreds of native pilgrims on top of Sri Pada – 'Adam's Peak' – the sacred mountain top of Sri Lanka – or by the social links of belonging created around a campfire while singing 'Waltzing Mathilda' in the Australian outback. Such a place is remembered very positively. However, Tuan (*ibid.*) does not stress formative experiences but rather tries to sort out the factors that influence them. From the human point of view, familiarity, culture, mythology, ideals and good health are formative factors whereas, from the point of view of nature 'environ-ments of persistent appeal' can be such places as seashores, islands, valleys, a well tended countryside, a wide view or a historic cityscape. However, not all place experiences are pleasant: **topophobia** is repulsion of place. These can be 'land-scapes of fear' (Tuan, 1980) – places we might find threatening, such as barren heaths in horror stories or slums in which we feel we are intruders (see the extract from bell hooks at the beginning of this chapter).

Entrikin (1991) argues that the concept of place does not fit standard categories that separate objective and subjective approaches and divide the particular from the general. Place has to be regarded as having both the objective characteristics of location as well as the subjective experiences of the individual. When writing about places we have to 'consider both the particular features of localities and the generality of the idea' (Relph, 1996, p. 913). When introducing the idea of **action theory** (Werlen, 1993 – see p. 148), we need to employ yet another approach to place, this time focusing on how places change as the result of the outcomes of intended actions.

In these processes of change there exist tensions between the particular (or local) features and general (even global) processes. Relph (1976, p. 913) suggests

that the local and specific aspects of an environment enable us to discuss somewhere as '**place**', whereas the non-local international and general influences can
be regarded as creating '**placelessness**':

> It is . . . misguided to treat these as two separate phenomena, or even to see them
> necessarily as being in conflict. Rather, they are each implicated in the other, the local in
> the non-local, the general in the particular. In some contexts, such as old villages with
> traditional cultures, the particular qualities of a locality dominate and placelessness is
> subservient; in other cases, such as airports, standardized design prevails and the spe-
> cifics of place are scarcely discernible. Whatever is placeless helps to make places com-
> prehensible to outsiders.

Although **placelessness** may be given some 'positive' connotations as a concept,
placelessness is, on the whole, a polemic weapon connected with the **postmodern
critique** of modernist architecture and planning (see Box 5.4). Whereas Relph
mainly connects the identity of place and placelessness to **physical features**, we
need also to connect general, placeless trends and local, place-specific features to
the realms of **social relations** and **meaning**. As was demonstrated in Box 5.3,
social relations such as those between men and women as well as those between
workers and capital owners are influenced by increasing 'placelessness' –
although they do retain some place-specific features. Sack (1992, p. 92) points out
that 'A group's commitment to a particular place requires that it be infused with
shared meanings, and the strongest means of doing this is through myth, ritual, or
religion, because within this intense mode of experience, symbols become conflated with what they represent'. Although mythical and religious experience is
personal, it is also a shared experience that is reinforced particularly in static,
closed societies. Modern communications have diluted much of the **place-specific
realms of meaning** but there exist, even in the USA, strange pockets of 'otherness', such as the Amish lands in Appalachia, where people adhere to values that
disappeared elsewhere in the world before the First World War. Sack (*ibid.*, p. 93)
reminds us, however, that even in modern society there are attempts to infuse
national place with something like sacred meaning, but these efforts are diluted
by secularism and scepticism. The American dream (the 'land of the free'), which
was enforced through the establishment of national parks and was symbolized by
the idea of the 'self-made man', is not so 'sacred' that it can be ridiculed in such a
successful film as 'Forrest Gump'.

**Box 5.4 Placeless modernism and the postmodernist revival of the
place-specific**

In the 1880s, the invention of carbon paper unwittingly speeded up international trends
in architecture: façades and window frames designed in central Europe were copied,
hence making an impact on cityscapes all over Europe. Inexpensive mass transport
meant there was no longer a dependency on local materials for house-building: on the
treeless western skerries of Norway, houses had (until the 1880s) been low constructions with turf roofs and stone walls; a few decades later they had become replaced
by houses built in the new, national, wooden style of architecture.

The mass production of housing built in a grand, modern style was, however, primarily a feature of the 1960s and 1970s. The housing shortage that resulted as a consequence of the Second World War created a demand for quickly built, inexpensive modern housing. New methods of prefabricating walls (element building) made this possible, and large, uniform housing estates grew up all over Europe.

The most uniform and extensive housing estates were built under communist rule in the USSR and eastern Europe. From Budapest to Vladivostok, these housing estates all look alike, cheered up recently, however, by the odd, colourful, placeless McDonald's. The contrast between such a housing estate in suburban Budapest and a street in old Budapest gives a striking illustration of 'placelessness' and 'place', to use Relph's terms (Figure 5.4).

From the late 1960s in western Europe, reactions to the 'placeless' features of modern housing estates, offices and shopping precincts came into conflict with current trends in town planning: the rehabilitation of old houses became fashionable, rather than their replacement with modern, functional buildings. Hence romantic, postmodernist ideas about architecture and planning were to become an early attack on modernism.

Figure 5.4   Budapest – old and new, place and placelessness

There has, points out Massey (1994, p. 168), been a tendency to look upon the **identity of place** as something static, equating time with 'becoming' and space with 'being'. We may, however, look upon place as having been formed and re-formed over time 'out of a particular set of social relations which interact at a particular location. The individuality of a place is constantly changing, but it is still dependent on the particularity of relations that form a location. For instance, in many presentations of place as 'home', a secured stability and bound-edness is imagined, which most personal experiences tell us are debatable: 'The identity of place does not derive from some internalized history. It derives, in large part, precisely from the specificity of its interactions with the outside' (*ibid.*, p. 169). Such changes in relations with the outside can lead latent, local conflicts to explode, as was the case in Gilgit, Pakistan in 1988 (Box 5.5).

Box 5.5   The advent of a religious conflict in Gilgit, Pakistan

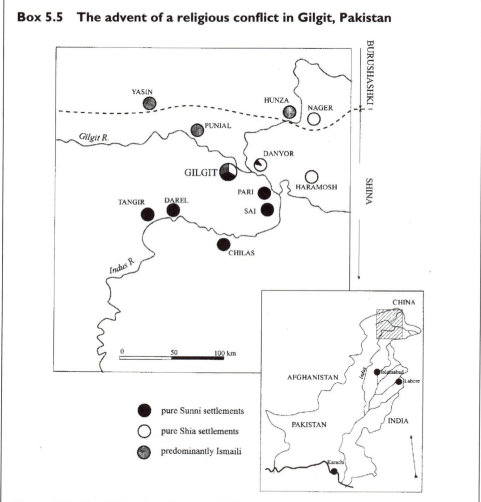

Figure 5.5   The Gilgit region of northern Pakistan
Source: Aase, 1999.

Until 1988, the Sunni and the Shia muslims in the Gilgit area of Pakistan (Figure 5.5) lived in harmony with each other, and intermarriages were frequent as people followed the tradition of *kaom* (a form of 'caste' division) rather than basing their marriages on religious status. In 1988, however, fighting broke out between Sunnis and Shias as a result of a quarrel over the start of Ramadan. Anti-Shia graffiti smeared on the walls in Gilgit on the second day of Shia fasting, and which claimed that Shias were *kafir* (heathen), triggered off shooting. This occasion has become a highly significant historic event in Gilgit; nothing is the same after the *tension*. Segregation amongst the groups is now a fact: intermarriages no longer happen. The puzzling thing is that the religious roots of this conflict date back to AD 660 and concern the rightful successor as the Fifth Caliph. If this strife has not caused serious trouble before, why then now?

'We have to go beyond religion in order to grasp the process of re-signification that has taken place in Gilgit. Symbolic processes are always contextual; likewise, the religious discourse in Gilgit yields no meaning unless it is situated in time and space' (Aase, 1999). In earlier times, the different religions of the region were fragmented into various independent valleys and settlements; the area was called the 'unruled country' in British colonial times. Each of the three religious groups (Sunnis, Shia and Ismaili) lived in separate principalities, and 'reasonable' mixing as a result of *kaom*-based marriages was not a problem. Some of these principalities retained their semi-autonomous status until 1974, when the northern region of Pakistan was divided up into four districts to be governed more effectively from Pakistan's capital, Islamabad. The northerners, however, felt they had lost local power and were being neglected by central government. This was especially the case with the Shias and the Ismailis, as the majority of Pakistani muslims are Sunnis.

In addition, Gilgit has developed as the administrative capital of the region and, as is often the case in growing towns, there has been a mixing of different religious groups in the town (see figure 5.5). Modernization and new forms of employment have led to a reduced importance of the *kaom* structure, and affiliation to a religious group has become a stronger means of identification. This has made Gilgit the most sensitive part of the region, and the town was the centre of the build-up of the 'tension'. The quarrel over the start of Ramadan was the 'straw that broke the camel's back' – the camel was already 'overloaded with goods from outside'.

While one factor in the conflict has been the change from local to central rule, another has been the opening up of the territory through the construction of the Karakoram Highway. This road has integrated the region into the national economy and has brought in new business as tourists and other foreigners have started to come into the area in growing numbers. However, this has not equally benefited all local groups: the Shias remain cultivators whereas the Ismailis are enterprising businessmen who have profited as owners of shops, hotels and bazaar stands in Gilgit. The Ismailis, on the other hand, have to compete with the Sunni Pathans who have migrated from the lowlands and who now operate one third of the bazaar in Gilgit.

Hence the Shias have been deprived of local political power and have benefited little from **modernization**. This deprivation, of course, has been worsening over a period of time, and to be called 'heathen' on smeared graffiti was to add insult to injury. Religious affiliation has served as a symbol for this growing tension, and it has become the dominating factor in the distinction between 'us' and 'them' (*ibid.*). This example shows how stable, local factors change when influenced by external structural forces, but in specified ways that make an understanding of local conditions crucial. This is what geography is all about. It also shows the importance of **political geography**, of power relations and of the concept of **territoriality**.

The positive image of a bounded, controlled place (rather than as place strongly influenced by outside relations) is also a **gender-related issue**. Hartsock (1983) maintains that the female sense of self is 'connected to the world', whereas the male sense of self is separate, distinct and even disconnected. Men make expeditions into the world, but belonging, identity and security are created by the bounded places they control: home, workplace, car. The female sense of self is founded on relations to others, particularly close family. Places are then meeting-places: Massey (1994) maintains that big city life, with its openness, its uncontrollable complexity, is more attractive to women than to men. This may also be regarded as an emancipatory move away from male-controlled bounded places, such as 'home' might be. The female sense of self is strengthened when women break out of bounded places and establish outside relations. Males react the other way round: they are already out in the world. This concerns power over place and introduces the idea of **territoriality**.

Box 5.6   Territories of a Sri Lankan tourist resort

A tourist hotel sited between palm trees on a sandy beach in a Third World country seems a 'Shangri-La' of wealth and affluence in a world of general poverty. Such a place attracts beggars, fortune-tellers, taxi drivers and youngsters selling T-shirts – all in the hope of earning big money. The hotel wants to protect its guests from these 'fortune-seekers', partly because it does not want competition on the territory reserved for the hotel shops. So the hotel's territory is guarded as a place only for the guests, hotel employees and entertainers who have been invited in by the manager. Even the native bus-drivers who transport the tourists from hotel to hotel have to sleep in poor accommodation in a part of the building that does not allow them access to the hotel's facilities. Between the palms, guards are posted to keep the beggars out, while inside fat middle-aged Europeans are served drinks at their beach chairs.

The hotel is a pleasant oasis but, at the same time, a tourist 'ghetto' the guests do not dare break out of. When they do so on foot through the main entrance, they are at once offered taxis and guided tours and have to break through a phalanx of other offers and appeals for help. If the tourist manages to refuse all these requests and walks a couple of hundred metres from the hotel, he or she will find him- or herself in another territory. Here there might be a small child or two asking for a 'pen' but, principally, the tourist will be left to look around, to sit undisturbed on the balustrade or to walk around the marketplace.

The ring of fortune-seekers around the hotel is also a territory, and this might be divided up into smaller territories. Each beggar (or family group) has carved out a definite small territory of his or her own. There might be competition over boundaries and zones of ambiguity, but the beggars, pimps and prostitutes have territories of action they try to control. Only those with a great deal of self-confidence and strong personalities can afford not to bother about these territories and to trespass anywhere.

Most of us need territories we feel we can control. A growing child will constantly try to enlarge his or her territory, but the territory the child feels is safe, and that which is unsafe, is always at the back of the child's mind.

TERRITORIALITY AND 'OTHERNESS'

A cultivated field is a place but it is even more an area that needs ploughing, weeding, and cultivation if it is to remain a field producing what its owner has decided it should produce. A cultivated field has this form because all other possible uses are not permitted in it. Virtually all places in the human world are such controlled places. When we control areas in space we create **territories**, and the act of creating territories is the geographical exercise of power. At the same time, territoriality creates and sustains real places, such as the living-room, the home, school, as places of specific functions and thus of meaning (see Box 5.6).

Positionality is a basic source of our understanding about the world, and of our ability to impose territorial power within it. Thrift (1983) points out that positions can be either **compositional** (what sort of person you are – based on class, gender and other characteristics) or **contextual** (where you come from – an insider or outsider in a place) or a combination of both. Powerful groups (people in dominant positions within a society) impose their interpretations of place, landscape and nature on society at large, and will generally define territorial boundaries. These interpretations and boundaries may, however, be challenged by other, often marginalized, groups in the society (see Box 5.5).

Geographical discussions have focused on conflicts between the **core** and the **periphery**, between **industrial** and **developing countries**, between **multinational companies** and **small-scale, local firms** in the process of the **globalization of the economy**. It is impossible to cover these important geographical themes adequately here, and so the reader is referred to Peter Dicken's book *Global Shift* (1998, third edition) and Allen and Massey *Geographical Worlds*, 1995. To exemplify our discussion on territoriality within geographical research, we will focus here on **social exclusion**.

Johnston (1997, p. 289) points out that **postmodernism** emphasizes the existence of the **Other** in society – groups 'whose identity is defined by the dominant groups'. Dominant ideas (often prejudiced, stereotypical opinions on race, religion, class, gender and nation) are powerful sources of human identity and a means of establishing territorial boundaries or legitimating the use of power to protect vested interests. Gregory (1994) uses as an example the Gulf war of 1991, the first 'postmodernist war' in which planetary television networks projected an Americo-centric vision to an audience that lacked any alternative grid, for example on the legacies of colonialism. In order to rescue Kuwait (which could be regarded as a puppet state created by multinational oil companies), media presentations made Americans able 'to believe the worst about our "enemies" and the best about ourselves' (Stam, 1992, cited in Gregory, 1994, p. 205). Gregory (1994) says that the point here is not whether we agree with the Gulf war or not but the question of authorization, on whose terms is a space created in which 'they' are called upon to speak?:

> Most of us have not been very good at listening to others and learning from them, but the present challenge is surely to find ways of comprehending those other worlds . . . without being invasive, colonizing and violent . . . we might come to realize not only that our lives are 'radically entwined with the lives of distant strangers', but also that we bear a continuing and unavoidable responsibility for their needs in times of distress.
>
> (*Ibid*, p. 205)

The third quotation at the beginning of this chapter (Foucault, 1980) reminds us that power relations give meaning to space and infuse empty space with **meaning**

as territories. Feelings of ownership and belonging attach to **national territories**: these contain national heritage, history, traditions of social relations and religion, language, and shared symbols of physical and ethnic identity: 'The myth of cultural homogeneity is needed to sustain the nation state, to ensure support for domestic and foreign politics which are conducted on behalf of the state' (Sibley, 1995, p. 110).

A definition of national culture requires a contrast – a definition of **'otherness'** that is seen as a threat to national identity. However, the idea of a purified national identity sits uneasily with the flows of people and cultural fusions that are generated by global capitalism: cheap labour immigrants from the Third World have played an important role in the growth of wealth of industrial countries. Illegal Mexican immigration to California is a typical example. As documented by Kearney (1991), the US government's policy is not to keep illegal immigrants out but to regulate their flow, so that the immigrants' ethnicity can be dissolved to make them 'Americans'.

In Norway, Pakistani immigration to the capital (which is not large by international standards) has created political conflict. The reason is that the Pakistanis do not 'become Norwegians fast enough': they stick to the Muslim religion and have even built mosques; the women do not learn Norwegian and stay in their homes; they have more children than average; and they create 'learning problems' for 'real Norwegian' children in those schools where their children are most numerous. While cultural differences are not acceptable to the majority, racial differences are, as witnessed by the integration of children from Third World countries who have been adopted by Norwegian families. Political pressure has forced the Norwegian government to introduce strict immigration rules, although individual humanitarian actions and 'church refuges' have prevented some from being sent back to their home countries. Immigration policy-makers in Britain and Germany now have to adhere to EU policies which, however, are creating a 'Fortress Europe'. This denies access to poor people from non-EU countries and questions the place for many 'already there, providing cheap labour for core industries' (Sibley, 1995, p. 110).

Since the 1930s, the geographical investigation of ethnic and racial segregation in cities has been a research theme of American urban geography and sociology. The roots of this research can be traced back to a study of the social space of the black community in Philadelphia, carried out by DuBois (1899). DuBois's study was empirical, being based on statistical and graphical description, but it was also complemented by detailed case studies. Interviews based on formal questionnaires took up such themes as crime and job discrimination, and they documented the territorial enforcement of housing markets:

> The undeniable fact that most Philadelphia white people prefer not to live near Negroes limits the Negro very seriously in his choice of a cheap home. Moreover, real estate agents, knowing the limited supply, usually raise the rent a dollar or two for Negro tenants if they do not refuse them all together.
>
> (*Ibid.*, pp. 295–6)

The influence of bad homes in bad neighbourhoods, combined with poor access to work, created a situation of **social exclusion** which made it hard for blacks to retain self respect and ambition; rather, DuBois noted, it bred carelessness and

criminal behaviour. Sibley (1995) points out that DuBois's work was rather neglected by the dominant Chicago school of sociology, whose leading figures were
Robert Park and Ernest Burgess. The Chicago school had a strong influence on
spatial science work in urban geography. Its members discussed models of diffusion in urban concentric or sectorial zones and observed that, as the economy and
job opportunities changed, the southern 'Negro' was seen to climb up the economic ladder.

Similar studies were undertaken in Europe, but these concentrated on not so
much racial as cultural differences. The conflict in Northern Ireland has, of course,
been a focus of interest in the UK. One example of this was the mapping of
different patterns of grocery purchases made by Protestants and Roman Catholics
in Belfast (Boal, 1969). This study showed that Protestants principally shopped in
precincts in the eastern part of the city, while Catholics shopped in the west. Such
studies were based on urban statistics, not on qualitative research that seeks to
understand the actions of 'others'. The newer approach, which may be termed
'the cultural turn' in geography (influenced by postmodernism and feminism, in
combination with Sayer's suggestions for 'concrete, intensive research' (see Figure
4.6, p. 131), is just such a call to 'go out into the world and experience the life-
worlds of other people' (Box 5.7).

Box 5.7   The revitalization of cultural geography

Cultural geography, as presented by Sauer and the Berkeley School (see Box 2.9, p. 52),
focused on the interpretation – through fieldwork – of the material aspects of the
cultural landscape. To explain cultural patterns, one would investigate the customs,
traditions and innovations associated with the culture of certain human groups (e.g.
'hispanic' Americans). Culture is hence seen here as the property of human groups, not
individuals. By basing their explanations on the notion of culture as supraindividual,
Jackson (1989, p. 18) notes that 'cultural geographers have often failed to address the
wider social context in which cultures are constituted and expressed'.

A new 'cultural trend' in geography is now well established, as this chapter has shown.
Much research is focused on individual 'maps of meaning' in humanistic studies, or on
the social relations and cultures of subgroups.

Jackson (*ibid.*, p. 180) sets a programme for this new **cultural trend** by arguing that
culture is

the medium or idiom through which meanings are expressed. If one accepts the preceding
arguments for a plurality of cultures, then 'culture' is the domain in which these meanings are
contested . . . studies of women and minority groups . . . must now be superseded by studies
of gender and racism which focus on the relations between dominant and subordinate groups.
Similarly, cultural geographers should not ignore dominant cultures where a whole new agenda
remains to be investigated. Recent work on nationalism, monarchy, and heritage, for example
. . . suggests a range of avenues for future research, whether or not this involves a conception
of 'class culture'.

A number of geographers have argued for the significance of locality in the
quality of everyday life, but there has also been debate over of the importance of
the local in modern societies. Many have argued that the traditional *Gemeinschaft*

Figure 5.6 Definitions of neighbourhood in Cardiff by (a) activity patterns and (b) perceptions
Source: Herbert and Raine, 1976, pp. 328–9

society (which characterized rural villages and implied tightly knit patterns of primary relations in socially relatively homogeneous communities) has disappeared, to be replaced by an urban type of *Gesellschaft* society in which individuals participate in formal, role-directed relationships with others who do not necessarily live nearby. Modern **network studies** of active middle-class people confirm this: people have relations with others in the same trade or in clubs of shared interests (the golf club, health studio, rotary association, dance club, etc.). The growth of *Gesellschaft* society is a strong tendency in modern society just as is increasing 'placelessness', but still we find that local connections are important – local *Gemeinschaft* society has not yet died out.

This has been pointed out particularly by female writers looking at 'kitchen-table societies' as experienced by women, children and elderly people who are dependent on local networks for their everyday lives. It has been shown that, although people move out of their home environments much more frequently to work, shop, etc., a great many of an individual's needs are still and, in many cases, have to be, met in the local environment. These types of needs can be described as follows:

1) Psychological (provision of food, shelter and health care).
2) Safety (privacy and protection from physical harm and intruders).
3) Affection – belonging (identification with, acceptance by and relation to other members of the community).
4) Esteem (status and recognition by others).

5) Self-actualization (creativity and realization of one's potential).
6) Cognitive/aesthetic (intellectual stimulus and experiences) (Walmsley and Lewis, 1993, p. 224).

'A place may be a locus of access to relational resources. Living in a place exposes people to particular processes of constraint and opportunity which significantly shape their social worlds' (Healey, 1998, p. 58). We all try to protect our access to such relational resources, most often by asserting some sort of territorial control. Geographical research has in this context focused very much on the concept of the **neighbourhood**. Town planners have used a 'neighbourhood principle' when designing new suburban housing estates. Such a neighbourhood should be closed to fast through traffic and should include a separate pedestrian network so that children can walk safely to school or kindergarten or to a bus stop or the local shopping centre. Recent geographical and sociological research has, however, focused on the difference between **perceived neighbourhoods** and planned or 'real' neighbourhoods. In a study of Cardiff by Herbert and Raine (1976, see Figure 5.6), residents in three streets were questioned about their definition of the local neighbourhood and, at the same time, were asked to describe their activity space concerning local friendship patterns and their use of services. In district A there was strong correlation between perceived home area and patterns of social interaction. This was, to a lesser extent, the case in district B, but not at all in district C. One reason may have been the physical layout of each area: for instance, a lack of local service institutions in district C or the provision of local-level social structures. Planners or city managers tend to see the city as a collection of segregated neighbourhoods. However, from our understanding of the ways individuals and groups of people use the city on a daily basis, the city can be seen as a collection of **overlapping neighbourhoods**. Research on people's perception of neighbourhood shows major differences in different people's spatial activities according to their age, gender, class and ethnicity (Madanipour, 1998). A closely knit neighbourhood may involve positive inclusionary, as well as negative exclusionary, practices. We will now look at the geographies of exclusion.

GEOGRAPHIES OF EXCLUSION

Ongoing geographical research into the 'geographies of exclusion' sums up neatly the different 'cultural' trends in human geography. However, before discussing these theoretical links, we will give some examples of exclusionary practice as presented in empirical research.

The **indoor shopping centre** has developed as a significant new mode of retail service provision which, at the same time, provides a new fantasy world of leisure and consumption. In the UK the Meadowhall Shopping Centre in Sheffield 'recreates the romance of Paris and Florence under one roof' (Sibley, 1995, p. xi); the Metro Centre in Gateshead even includes an indoor amusement park. Such places have created a new form of holiday experience: pleasure combined with the 'needs' of shopping. In appearance these centres are 'classless' – a consumption paradise everybody can enjoy even if, in reality, they are not 'classless'. They are located in the main out of town so that shoppers have to come by private car or pay for public transport. The centre's management will proudly announce that the centre is for all the family, preferably with a credit card, and that everything has been done to ensure the centre is a safe and clean environment the family can enjoy. The management may point out that the centre was located out of town so that idle youngsters, vagrants, etc., will not put in an appearance; if they do, the centre has guards to evict them. This suggests, says Sibley (*ibid.*), that shopping centres like these constitute a kind of ambiguity: seemingly public, but actually private space.

A society's degree of **spatial exclusion** practices seems to be inversely related to the degree to which that society may be regarded as egalitarian. In Scandinavian countries (in which historical tradition, welfare provision and job opportunities until recently tended to level off class distinctions), it is difficult to distinguish between upper middle-class, lower middle-class, working-class and social housing. Scandinavians who study housing provision in the UK and, even more so, in the USA, experience a shocking new world of difference. A simple explanation for this is, of course, the traditional role of class in UK, and the ethnic diversity and the 'self-made man ideals' (by which poverty is your own fault) of the USA.

The white **American suburb** stands out as a prime example of exclusion in practice. Some of these suburbs have been described neatly as 'enclosed communities', socially purified and defended, fortress-like, against the supposed threat of the poor. North of Los Angeles, the author visited such a suburb – located on a hilltop around a golf club with no access other than a road for private cars. In the south of Los Angeles I encountered people living in cardboard boxes on the sidewalk, or as squatters in closed-down office buildings – a locality in which I did not even dare to stop the car. In Canada there is an ongoing discussion about planning regulations in areas designed for low-density housing – about whether extra apartments for rent should be allowed in houses in such neighbourhoods. The argument goes that allowing people to hire small apartments will devalue the neighbourhood's social standards, hence lowering the market value of adjoining houses! (Poulton, 1995).

In Europe, rising unemployment has led to increasing social problems, particularly in inner-city tenements and peripheral social housing estates. In the 1960s and 1970s in the UK, considerable 'brick-and-mortar subsidies' were provided by the government to build public housing estates, which accommodated some 31% of households by 1971. By the 1980s and 1990s general housing subsidies in the UK had, as in most other European countries, changed to personal subsidies for those in special need. Through the 'right to buy' and urban renewal, a large part of the UK public housing stock has been privatized. This has meant a spatial concentration of the remaining public housing, which now mainly caters for those who are unable to cope in the housing market.

In such **social housing**, there tends to be a concentration of lone parents, the long-term unemployed, socially disabled people, drug addicts, etc. This spatial concentration means that disadvantaged groups such as these are effectively excluded from relocating to other neighbourhoods. Recent research into housing conditions in the north of England has attempted to give voice to people living in these conditions, to explain how their basic routines of everyday life are carried out. Lone mothers explain how they have to navigate a route to local shops through streets that are controlled by gangs of young men and grown-ups who solve all conflicts physically; their children are sometimes found playing with

discarded syringes: children are hence kept inside the home and as a consequence, suffer from illnesses more than is usual for their age. A 'wrong' address
means it is almost impossible to get a job. Social and spatial barriers conspire to
make it almost impossible to attain the normal life expectations of acquiring a
home and a worthwhile job; young people are kept in 'their place' (Speak *et al.*,
1995).

Healey (1998) suggests that the meaning of **social exclusion** and the significance of place and neighbourhood could be discussed in the terms of the following concepts, which are closely linked to newer trends in human geographical
research: webs of social relations, structure and agency, sociospatial relations, and
everyday life and neighbourhoods as living places.

'A *relational view of social process* takes a view of individuals, not as autonomous
subjects with individual preferences, but as formed within social contexts' or
through social networks (*ibid.*, p. 55, emphasis added). For the majority, place-
based *Gemeinschaft* networks have, to a great extent, been replaced by placeless
Gesellschaft connections. Social networks intersect and cluster in particular places,
but these are not necessarily residential neighbourhoods. The main point here is
that **relational space** is of crucial importance in the active processes of building
relationships and meaning, in social exclusion and in cohesion – which are active
social processes. Research indicates that neighbourhood relations (for good and
bad) are more important for those who are outside 'mainstream society' than for
those who are actively involved in the workforce and cultural and political life
(see Box 5.8). Strong social cohesion in a disadvantaged neighbourhood might
give people strength and identity, but this is not necessarily a good thing: it might
reinforce the inhabitants' social distance from the rest of the society.

Structure and **agency** are two concepts related to Giddens' structuration
theory. Past patterns of behaviour carry power forward to the present in the
micropolitics of what we do, and these – our actions – have the potentialities to
change gradually structuring dynamics. Structuring forces generate fractures and
divisions which, in the UK, have their roots in the language of class and distribution. The division between working class and capitalism is inherited, but in a
postindustrial society this division has led to a situation where exclusion means
exclusion from mainstream society: three quarters of society belong to the 'mainstream'; one quarter of the population are economically, socially and politically
disconnected as this sector of society is outside the workforce, is not involved in
social community work and, to a large extent, does not vote. In our research into
deprived neighbourhoods, therefore, we have to look for other structural forces
and relations between structure and agency than those based on traditional Marxist ideas. Society has changed, and so have the structuring forces.

Sociospatial relations are the main focus of human geographical research, and
an understanding of how a sense of place is an important dimension in people's
everyday life is essential if we are to deal with social exclusion. Certain living
places are excluded from normal life as these are places of divergent behaviour;
places with, for instance, high levels of crime. In building up a neighbourhood's
organizing capacity – building up positive sociospatial relations – it might be
possible to draw people who are suffering from social exclusion back into mainstream society. In practice this means creating jobs or empowerment through

---

Box 5.8   The concept of social exclusion

Globalization has reshaped the social and spatial geography of cities, invoking a growing social divide. Long-term unemployment, particularly for manually employed men, has undermined the quality of life for an increasing sector of the population. There has been an attempt to describe this situation by dividing the population into quarters: the first has a secure income, a good life and does not bother too much about the future; the second has a reasonably good life but is concerned about the future; the third lives with insecure job possibilities and has problems in meeting the requirements of daily life; and the last is excluded from participation in the job market and exists outside mainstream society.

Madanipour (1998) points out that, as a concept, *social exclusion* suffers from lack of clarity, especially as there is a degree of ambiguity between poverty and social exclusion. The constitution of social life is such that different forms of exclusion are fundamental to any form of social relationship. The division between public and private spheres, for instance, means drawing boundaries around spatial domains and hence excluding some from these spheres. Exclusionary processes also work in close partnership with inclusionary activities to maintain society's fabric. We can, however, identify 'economic, political and cultural arenas as the three broad spheres in which social inclusion and social exclusion are manifested and, therefore, can be analysed and understood' (*ibid.*, p. 77).

In the *economic arena* inclusion is access to resources, mainly via secured employment. Exclusion in economic terms is thus the lack of access to the labour market. In the *political arena*, inclusion means having a stake in power, being able to participate in decision-making. A lack of participation in organizations and refraining from voting in elections stand out as characteristics of exclusion in a political sense. In the *cultural arena*, the main form of inclusion is the sharing of a set of symbols and meanings, such as language, religion and nationality. Forms of cultural exclusion vary widely: 'as experienced by minorities whose language, race, religion and lifestyle are different from the larger society' (*ibid.*, pp. 77–8).

In its most acute form, however, social exclusion appears to occur when factors of economic, political and cultural exclusion combine and dominate certain localities. Such combined concentrations of disadvantage are particularly apparent in inner-city tenements and peripheral social housing estates. This localization of social exclusion is an important concern for social geography.

---

participation in community governance. To overcome the adverse effects of social exclusion, research would profit from looking at the following questions:

> 1) how social worlds of place affect people in terms of mobilising for change (where do people look for allies, who are they mobilising against etc?); 2) what brings outsiders to seek to engage in place based mobilising in someone else's place; and 3) what conceptions of place and community are used by these insider/outsider mobilisation processes?
> (Healey, 1998, p. 59)

The concept of **everyday life** provides us with more focus to understand what living in places actually means. An everyday-life perspective focuses on the multiplicity of roles people play and the services they use each day. This is a bottom-up approach, in contrast to a top-down research approach that might study the efficiency of service deliveries in a particular area. This bottom-up approach takes its inspiration from the postmodernist scepticism about general models of explanation: humbly, one tries to let people speak for themselves, to elucidate the meanings, metaphors and ideas they follow in their day-to-day living practices.

Neighbourhoods are places of multiple motion, and the concept of **neighbourhoods as living places** focuses on this. Recent research has shown that, for good or bad, for people in deprived neighbourhoods these are the key places that define their social world. A **realist** approach that focuses on the relationships between the necessary and the contingent may find that necessary structural relations are not as important for such people as individual, contingent relations and roles in the local living space. We can thus speak of 'my place', 'your place' and 'their place'. Cruddas Park in Newcastle upon Tyne was both 'the place I call home' for an insider and 'an appalling place' for an outsider (Wood *et al.*, 1995): 'An important dimension of neighbourhood improvement work is to consider how people value their living places, how they compare them with others, and within what social context this valuing takes place' (Healey, 1998, p. 62).

The '**cultural trend**' in human geographical research (Box 5.7, p. 169), which has developed since the late 1980s, focuses on difference within society and on such contingencies as ethnicity, gender, nationality, political persuasion, etc. It also involves distrust of 'grand theory'. Different approaches to qualitative research have been tried out as geographers have started to investigate ordinary people's everyday lives, much in the same way as planners have advocated a bottom-up approach to the planning process to involve citizens in **social mobilization** (Friedmann, 1992). Johnston (1997, p. 271) maintains, however, that the 'cultural turn' lacks a definite core. The relative peripheralization of the spatial scientist is, however, a fact that has occurred in parallel with the growth of qualitative research and an interest in everyday life.

At the same time, in most geography departments there has been a growing interest in **GIS** (geographical information systems) as an aid to research. As GIS are increasingly taught in MA and PhD programmes, this might lead to a new interest in more quantitative approaches. We might predict, therefore, that an increasing number of young geographers will try to find out what new knowledge and perspectives GIS will bring to geographical research. So far the problem has been to find teachers who can teach the next generation the data operations required in, and research possibilities offered by, GIS. At the turn of the millennium we will find geography departments armed with both appropriate computer capacity and the researchers able to utilize GIS – it would be very strange if geographers did not take up the opportunities offered. It is therefore appropriate to discuss the possibilities of GIS as well as the societal trends that might lead to new geographical research projects.

GEOGRAPHICAL FUTURES

As the geography of our world changes, the discipline of geography will change in response to new challenges. Haggett (1996) identifies seven trends in world geography, and I present below his personal reflections on the effects these trends will have on geographical research. The seven trends can be summarized as follows:

1) The world's population will continue to increase, but a much larger part of it will be elderly people. This increase will primarily come from the Third World, in which there will be an explosive growth of the urban population.
2) If the Third World is allowed to increase its consumption of energy and other natural resources up to the average consumption in industrial countries, global resource demand may be increased threefold by the year 2025.
3) The combined effects of the two first trends will be increased environmental pressures at both local and global levels.
4) There will be a continuing switch in resource extraction to offshore areas, thus creating possibilities for new types of international dispute.
5) The world will become increasingly 'smaller' – with reduced travel time, increased travel opportunities and the rapid improvement of information technology.
6) The electronic revolution is just beginning. Power and influence are likely to come about through access to networks, and long-established hierarchic structures are likely to be replaced by ring structures.
7) The weakening of hierarchic structures might have implications for geopolitical power. The electronic revolution and the globalization of the economy have weakened tremendously the power of nation-states. A former super-power (USSR) has collapsed, and the USA is in a relatively weaker position than it used to be (although it is the home of Microsoft). Whether the world will be ruled increasingly by big multinational consortiums that provide information and consumption or by rising new powers (e.g. the EU, China, India) remains to be seen.

Changing hierarchic structures, the impacts of **globalization** and the modernization of society are the main sources of inspiration for current geographical research. So far in this chapter we have mainly discussed geographical research that is concerned with the local effects of general processes. This emphasis on changing life-worlds in local places and on comparative case studies will probably remain a main focus for social geography into the future, and economic geography will probably focus on the local and international effects of globalization and the electronic revolution.

The pressure on global resources and the human impact on the ecological balance, both locally and globally, will, it is hoped, inspire both physical and human geographers to take up research within the field of **ecogeography**. Haggett (*ibid.*) believes there could be a breakthrough in climatic modelling – a synthesis that would bridge the gap between detailed meteorological studies, geophysical models and Pleistocene history. For instance geographers have for years been working to find possible links between tropical deforestation and climatic change – perhaps *one* reason for the desertification of the Sahel belt in Africa (see Box 3.5, p. 90) and links between climatic change and the decrease/increase in the growth of glaciers. Progress in this type of research will be aided greatly by improved computer capacities and time-series data. Haggett (*ibid.*)

points out that a global land-use inventory could be completed and maintained, using remotely sensed data and an international network of benchmark sites.

Haggett (*ibid.*) even prophesies a coming of age for **spatial analysis** based on the fact that spatial data are now easier to analyse in a secure manner than was possible in the 1960s. This is partly the result of changing possibilities in map production, which are 'likely to continue to change from a ROM to a RAM format' (*ibid.*, p. 969). ROM (read-only memory) format means conventional map production where displays of information are in a fixed format; for example, a map library with its printed sheets can be likened to ROM. Storing map information on computer, however, allows mapping data to be in an interactive, on-demand form – random-access memory (RAM), which presents only the specific information we need. GIS technology also makes it possible to extract quantitative data on area quality, terrain steepness, etc., from maps and satellite photos. We need, therefore, to analyse briefly the possibilities of **GIS as a new tool for geographic research** particularly in the field of **ecogeography**.

GEOGRAPHY AS HUMAN ECOLOGY

The subject-matter of **human ecology** concerns itself with those aspects of ecology that impinge on humanity's biological nature and humanity's relationship to the natural environment and resources. When we take up the study of **ecogeography** (or geography as human ecology), we need to be familiar with the tools of modern biological ecosystems analysis (Box 5.9).

Stoddart (1967) maintained that ecosystems analysis gives the geographer a useful tool but emphasized that model-building as employed by biologists cannot be carried over directly into geography. Chorley (1973a, p. 157) considered that 'the ecological model may fail as a supposed key to the general understanding of relations between modern society and nature, and therefore as a basis for contemporary geographical studies, because it casts social man in too subordinate and ineffectual a role'. When humanity began to clear forests and plough the land, the biosphere was changed from a system working in itself and for itself into a **resource** for just one species. Humanity became an ecosystem manipulator in order to develop those resources it was interested in to their maximum availability. Chorley (*ibid.*) suggested that the ecosystems biologists study are characterized by negative feedbacks, which almost always start to work as soon as something upsets the system's stability. Nature often tries to rectify the instabilities introduced into a system as a result of human intervention, but humans often manage to alter these adjustments or they may ignore them altogether. For example, when wheat cultivation in the drier areas of the North American prairies was about to result in a negative natural feedback (in the form of drought and dustbowls), the farmers – instead of giving up agriculture and allowing the ecosystem to return to its original balance – continued to grow wheat – but now, however, in a strip system every other year. Humanity very often succeeds in replacing a natural ecosystem with an artificial one and is able to learn from its mistakes and hence make adjustments.

Chorley (*ibid.*) maintained that the ecosystem model would be of geographical significance only in so far as humanity can be said to function in the same way as

Box 5.9   Ecosystems analysis

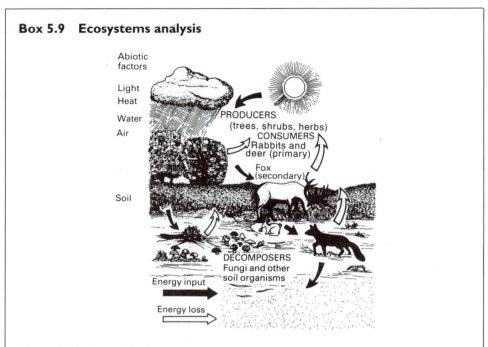

Figure 5.7   A generalized ecosystem
Source: After Bliss *et al.*, 1979

An **ecosystem** consists of the biological community at a specific place and the environmental physical circumstances that influence and are influenced by that biological community. An ecosystem is not just the sum of the flora and fauna, water, air and soil, for the most critical elements are those that circulate within the system: incoming solar energy, the transport of water and gases, the production of organic matter from inorganic material, and growth and movement. An ecosystem may be defined for any and every geographical region of study: from a small puddle to the globe itself.

A significant development in systems ecology has been the creation of operational models of circulation; models that imitate the relations between the parts of the system, including the means of subsistence. Stoddart (1967) points out that the concept of the ecosystem has a number of elements of interest to geographers. First, it is monistic: it brings together the worlds of humanity, animals and plants within a single system where the interactions between the components can be analysed, to a large extent, by quantitative methods. By emphasizing the functions and characteristics of whole systems rather than any particular relationship within them, ecosystems analysis bypasses a discussion of determinism and the conflict between physical and human geography.

Secondly, an ecosystem is structured in a more or less orderly, rational and understandable way, which makes it possible for other scientists to test an ecosystem model once it has been developed. Thirdly, an ecosystem's functions contain a permanent throughput of material and energy in processes that can be analysed and quantified. Finally, an ecosystem is a general system that uses the theories developed in **systems analysis**.

other species; it loses its significance when human beings succeed in manipulating nature. The use of the ecosystem concept in geographical analysis raises two important questions: are human systems so much more complex than the biological parts of ecosystems as to limit seriously the model's geographical use? If so, to what extent does humanity control the system within which it lives?

To the first question, a biologist would answer definitely not – artificial systems are much less complex than natural ecosystems. In its manipulation of nature, humanity always simplifies nature. In biological terms, because of the dominance of one species in a Mid-West wheatfield, the artificial system is much simpler than the natural prairie it succeeded; and an important ecological observation is that simple ecosystems are much more liable to negative feedback than complex natural ecosystems. This problem of negative feedback is also the case with the town – surely the most manipulated of all ecosystems. Here simplification creates a great many 'dead ends'. Sewage, for example, is channelled into what would be called a dead end in ecosystems analysis – that is, it is channelled into a system that cannot transform and recirculate the waste organic matter and thus, as a resource, the sewage is lost. Most pollution problems are a result of humanity's manipulation and simplification of the natural ecosystem, so creating dead ends and, ultimately, serious negative feedbacks. In this way valuable resources are depleted. Humanity may be able to control the system but, if it wants to avoid serious repercussions in the long run, it would be wise for humanity to work *with* nature and not against it. The ecosystems approach has real value when we analyse the impact and consequences of human activities on nature at the local scale; it also has value when we analyse the global issues of humanity–biosphere interactions. We have, of course, to consider the human abilities of actions intended to manipulate ecosystems, but this is not the main problem in using biological analogies in geographic research. The main problem with such analogies is that ecosystem analysis in biology is, in general, concerned with local, **vertical connections**; human systems today, however, are dominated by **horizontal connections.** This makes it difficult to assess the effects human actions have on the biosphere, on which we all depend for life. We all depend on natural resources, but what we eat and the artifacts we use in daily life come from so many sources and parts of the world that it is difficult to ascertain their original natural connections. When you buy garden furniture made of teak you may be contributing to the destruction of tropical forests; however, the retailer will probably tell you that these chairs were produced from teak plantations that are actually contributing to forest preservation.

We have been told to 'think globally, act locally', but because of the often great distances between the consumer and the location of resource extraction, it is increasingly difficult to comprehend the linkages. This is an obvious problem, but it is still no excuse for refraining from research into the effects of human activity on nature. Humanity–environmental issues are at the core of our discipline, and it is possible for both physical and human geographers to analyse at a local level the impact of human activities on nature. And the application of GIS, satellite photography and computer programs has recently opened up new possibilities for linking local impact analyses to broader, global trends.

GEOGRAPHICAL INFORMATION SYSTEMS (GIS) AND WAYS OF REPRESENTING THE WORLD

The technological revolution in mapping began with early satellite photos and was connected with **remote sensing** techniques, in the first instance developed in the 1960s and 1970s for military surveillance. When first introduced, remote sensing proved to have a great many advantages over traditional mapping methods. First, it simplified by spatial resolution (the size of the pixel to be used) the identification of objects that were to be placed on the map at a certain scale. Secondly, it allowed maps to be revised much more frequently and assessment of the urgency of such revisions. Thirdly, remotely sensed images could be used directly in the computer-assisted map-production systems developed in the 1970s and 1980s.

Different satellite sensors produce different images, and the range of these is expanding with each new satellite sent into space. Thousands of images are sent back to the earth each minute, and the usefulness of these images has long since been recognized as covering much more important ground than simple military surveillance. We now rely on remote sensing to a great extent for weather fore-casting, and images of changing vegetation patterns are an important means of monitoring ecological changes: 'remote sensing allows **resources** to be identified and measured at the same time as greatly aiding in the monitoring of environ-mental degradation', note Dorling and Fairbairn (1997, p. 106).

Developments in **GIS** were relatively unimportant until the 1980s. In 1982, Jack Dangermound launched the first successful application package, ARC/INFO. Dangermound (1992) maintains that the most important function of GIS technol-ogy is to provide one mechanism by which people of different organizations, different levels of government, different countries and different disciplines can come together to solve common problems. In this way GIS would serve the purpose of 'uniting the world'. It is true that GIS databases require large quan-tities of information which obviously need to be based on substantial co-operation. Dorling and Fairbairn (1997), however, express some doubt as to whether this 'coming together' will actually mean solving problems for those outside the ring of computerized power. Dangermound (1992) poses the idea that GIS may help in the development of an 'electronic democracy', as people in-creasingly gain easy access to databases held by governmental and other institu-tions. It is certainly the case that it is becoming increasingly difficult for any political, top-down organization to control access to data if it does not have overall technical control. Dangermound's argument, however, was made before Bill Gates founded the Microsoft Corporation which, in the 1990s, has created a monopoly that makes it very expensive to be technologically up to date and which may represent a new technical superpower.

So far, GIS is an expensive form of 'democracy', but this might only be a temporary setback: it is what GIS can do with data that is of most importance:

> In essence, some GIS hold their data as series of layers, and the creation of new informa-tion is achieved through the construction of new layers derived from the processing of one or more of existing layers. For instance, a layer of roads (represented as lines) could be used as an input into a process which output a layer of road corridors (represented as polygons 'buffering' those lines). . . . The cartographic possibilities of adding value to existing data through the use of GIS are not difficult to imagine.
>
> (Dorling and Fairbairn, 1997, pp. 128–9)

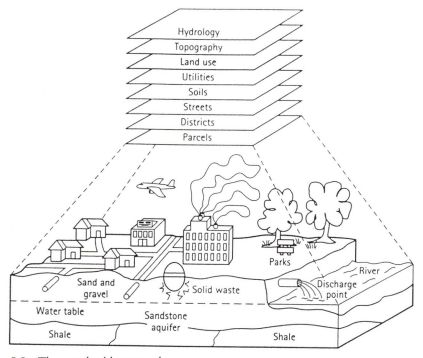

Figure 5.8 The standard layer mode
Source: Dorling and Fairbairn, 1997

In the geosciences, the three-dimensional display of geological and hydrological data is of great interest: GIS in these fields makes it possible to construct 'block diagrams' from contour maps and simple surface data, representing two-dimensional pictures of a three-dimensional scene (see Figure 5.8). By superimposing different layers of maps (including data on slope gradient, vegetation cover, amount of precipitation and exposure to the sun and dominant wind directions) it is now possible to forecast with much more accuracy areas that are prone to avalanches (Figure 5.8 demonstrates the principle of such superimposition). In human geography, demographic research using census data about the population living in each census district has made it possible to create extremely detailed maps of the social structure of cities. In addition to its usefulness in mapping, GIS can also be applied more generally as a medium of scientific analysis.

Openshaw (1991) maintains that GIS could resolve the identity crisis within geography. GIS will have a fundamental impact on the nature of the discipline,

> offering the prospect of reversing the disciplinary fissioning process and replacing it by fusion; a drawing together of virtually all the subdisciplinary products with their multitude of conflicting paradigms created over the last thirty years within a single (philosophy-free or philosophy-invariant or even philosophy-ignorant) integrating framework.
>
> (*Ibid.*, p. 622)

The technological research associated with the development of GIS is computer science, but the information it produces and the methods by which it can be manipulated are far more than yet another technical tool for geographers. As GIS offers a major change in the technology of mapping and the manipulation of map data, it is certain to provoke new geographical thinking and the development of new forms of analysis, see box 5.10.

Johnston (1997, p. 315) notes that the arguments of Openshaw and other GIS enthusiasts have been met with counterarguments that have clarified many of the 'differences between "social theory geographers" and "GIS-quantifiers" '. Social geographers argue that the **data** available for GIS manipulations do not just 'exist' as valuable sources of information; as they have been obtained for the most part for political, military and commercial uses rather than for intellectual ends, they tend to be biased. As pointed out by Dorling and Fairbairn (1997), there is no such thing as a 'free' data set. In much of the world outside the USA, where government data are freely available, data are copyright and large amounts even secret or confidential. The data that are available are, to a large extent, still concentrated in industrialized countries and, in particular, in North America. In Third World countries, on the other hand, most of the data needed for GIS analysis have not been collected. The nearest we get to free data is information available through World Wide Web, but this requires access to the Internet which is still dependent upon resources that are not available to universities in many parts of the world. And map information available on the web is rather haphazard, providing most data for the USA and almost nothing for many countries in the world.

Another social-science criticism of GIS is that it increases the level of surveillance of the population by those who already possess power and control. GIS concerns the precise location of land rather than people and, when people are mapped, it is to ascertain where they live and to obtain quantitative data on their living conditions. The more fundamental critique raised against **spatial science** (see Chapter 3) is also relevant to the debate on GIS. With this in mind, however, we should still accept that GIS offers geography a new powerful tool for description and analysis, particularly for physical geography and ecogeography. Whether GIS will be useful in, for instance, the prediction of environmental futures remains to be seen, but many of GIS's proponents uphold this as one of its aims.

Nordgård (1972, p. 29) says that, in the development of a science, we can distinguish three broad stages: descriptive, analytical and predictive. Description is the first and simplest step: it is concerned with the description and mapping of phenomena. The analytical stage moves a step further by looking for explanations and seeking the laws behind what has been observed. By the predictive stage, the laws have been studied so thoroughly that we can now use models to predict occurrences.

In the 1970s, geographers suggested that the discipline was moving into the predictive stage. Bartels (1973) presented a schema (Figure 5.9, p. 186) which, shows how human and physical geography were linked in the development of geographical research from its descriptive stage through the analytical and towards the predictive. Many researchers in physical geography and ecogeography will still find themselves working within the traditions presented at

Box 5.10 Geographical information systems (GIS)

A reading of any introductory text on GIS or of one of the many magazines devoted to GIS might promote the expectation that GIS can be used by almost anyone to solve almost any kind of problem. This is because GIS as a software product is heavily marketed as a tool for spatial data management, data analysis, data visualization and map production. However, users and potential users need to be fully aware of the limitations to the use of GIS as well as its capabilities.

The fundamental construct behind the current generation of GIS products is that the world is populated with features that can be described appropriately and discretely as one object: one of point, line, polygon or grid cell, or one of their derivatives. Many artificial features, such as telephone poles, roads and property boundaries, fit nicely into such discrete spatial descriptions. Many natural features or conditions (such as land cover) can be similarly described, although natural boundaries tend to have a fuzziness that requires special consideration. If features are determined through either digital air photographs or satellite images, the typical representation is as a homogeneous grid cell. However, data commonly associated with the social sciences may be more difficult to fit into a GIS framework without the sacrifice of some of their features. Such features as territories, neighbourhoods or families are delimited by imprecise spatial boundaries although they may be well defined in terms of **relational space**. Such objects are awkward to encode (without substantial loss or alteration in meaning) when employing the rigid GIS data structure, where geometric space dominates. Hence, while GIS can offer an alternative perspective to social science research, much research remains to be done before GIS can be considered a universal tool.

One typical GIS application is as a spatial database containing a rich store of information pertaining to environmental conditions. Planners, managers and decision-makers can draw upon these data to assess, monitor and predict conditions before, during or after a planned development (e.g. an oil and gas exploration), a natural event (e.g. a flood) or a human-induced disaster (e.g. an oil or chemical spill). There exist many implementations of such GIS in use at various levels of government and for different components of the earth–environment system. But these implementations all have one thing in common: the data contained in the database must be representative of the prevailing conditions. Without reliable, up-to-date information, any resulting decision would be suspect.

In the last decade or so, there has been a growing awareness of the need for improved monitoring of fragile Arctic areas in order to protect and conserve them, together with an improved understanding of the needs of conservation management, regulations and practices. In addition, there has been an acknowledgement of the growing threat on the quality of the Arctic environment from air- and water-borne pollutants from industrial activity in southern regions. This includes deleterious impacts on human health as well as on plant and animal life. Under the umbrella organization of the AEPS (Arctic Environmental Protection Strategy) (as signed by the eight circumpolar nations – Canada, the USA, Denmark/Greenland, Russia, Norway, Sweden, Finland, Iceland), a number of international and multilateral programmes have been established to conserve, protect, monitor and assess environmental conditions in the Arctic. An international GIS database is under development at UNEP-GERID Arendal in collaboration with AEPS projects to achieve a harmonized set of Arctic data to meet present and future assessment needs.

A glance through the associated web pages (URL http://www.grida.no/) shows the range of participating projects. However, the management of a project of this magnitude requires careful consideration and compromise in many different aspects of its implementation. In some academic institutions, issues of GIS management and construction are treated as research themes in their own right (URL http://ncgia.ucsb.edu/ncgia.html); elsewhere they are considered engineering or implementation problems that simply have to be overcome. To name but a few, these potential research issues include the following:

1) Meta-data, whereby data users are provided with adequate information about the source, quality and appropriate use of the data.
2) Data harmonization, or the combination of data from different sources that have different levels of generalization or different definitions (i.e. each country may use a different definition of 'Arctic').
3) The incorporation of traditional knowledge, or the integration of knowledge or lifestyle information from indigenous peoples, into a structured framework.
4) Co-operative spatial decision-making, where GIS acts as a focal point for discussion.
5) Data access or the distribution of data through technology (e.g. the World Wide Web).
6) Crossdisciplinary research (i.e. looking for relationships between ecosystem and/or societal components, such as environmental pollutants and health).

In the case of crossdisciplinary research, GIS itself cannot be used explicitly to answer cause–effect questions. At best, the co-location of numerous attributes can indicate a spatial correlation between the attributes that suggests a starting point for more detailed analyses containing alternative approaches. However, even the presentation of a view of what are considered to be by some researchers well understood themes can be problematic.

One of the data layers that would be useful in assessing any potential relationships between water-borne pollutants and the threat to health (human or ecosystem) is the one that contains information about ocean and nearshore circulation patterns. A cartographic view (such as found in an atlas) of such phenomena typically represents ocean currents by various sizes of blue or red arrows that indicate flow direction, magnitude and water temperature. In GIS terms, the arrow is a symbol rather than data and it needs to be converted (either explicitly or implicitly) to one of the standard data models – either raster or vector. And while the cartographic view presents a spatial and temporal generalization that can be interpreted easily for qualitative assessment, it is not *detailed* or *specific* enough for GIS-based analysis. Herein lies the problem.

The coastal ocean is a four-dimensional entity, with variation in *x*, *y*, *z* (depth) and time dimensions. This variation is continuous (i.e. the ocean never stops moving). The metaphor for the current generation of GIS is one in which map layers (variation in *x*, *y*) are stored as layers (either depths or time slices). To input such an entity into a GIS, a compromise must be made in order to deal with the mismatch between the complexity of reality and the data structure required by the GIS. Typically, data are generalized to present monthly, seasonal or annual averages. In the z direction, data are similarly averaged on to specified depth planes (e.g. I m, 5 m or 20 m depths).

While such averaging is necessary and relatively easy to carry out in a GIS, in both cases information about environmental processes acting at finer spatial or temporal scales is lost and therefore cannot be integrated into a GIS-based analysis. For example, seasonal averages do not allow the GIS analyst access to information about the process of spring-ice breakup in coastal waters. The presence or absence of ice could be a controlling factor on the distribution of oil or water-borne pollutants. And, unfortunately, it is often only after the database has been built and analysis has begun that the 'missing' information is identified.

What this means is that the GIS designer must work closely with domain specialists in order to ensure that appropriate information is included in the GIS database, and that it is represented in an appropriate manner.

Source: The material for this box was kindly supplied by Anne Lucas.

the lower end of this diagram. The predictive stage has, however, hardly yet arrived, and **postmodernism** has created doubt as to whether it will ever be reached. At the end of the millennium, optimism and a belief in modernity and science have been, for many scientists, replaced with doubt and insecurity.

FROM ORTHODOXY TO PLURALISM

A geographical synthesis remains as difficult to achieve now as at any earlier period in the history of the subject. Granö (1987) believes that the difficulties we have had in establishing a clear-cut **disciplinary matrix** may be due to the way in which the discipline was institutionalized by political decision in the universities during the late nineteenth century. It was only then that the first professors gave geography a cognitive content. They defended the infant discipline by trying in every way to emphasize the distinctions between geography and other subjects. With this in mind, the history of geography was extended backwards in time, with particular reference to its **cosmographic** roots. This was in line with the political expectations of governments and leading members of the geographical societies who had supported the establishment of geography at universities. At the same time the professors sought to prove their **scientific** abilities in relation to other scientists through advanced and specialist studies into such fields as glacial morphology and rural settlement patterns. The monistic unity of geography, observes Granö, was absolutely indispensable in order to justify its existence alongside other disciplines, but it did not suit the thematic, discipline-orientated structure of the universities. Geography itself gradually developed into a heterogeneous group of people comprising subdisciplines with no single theoretical framework, although they all maintained to aim at **geographical synthesis** in order to justify geography's separate existence. One of the most difficult problems in geography has been to create an exemplary scientific model to treat humanity–environment relationships.

For a long time the discipline-orientated structure of the universities functioned well. The institutions were small and they trained **specialists** to become univer-

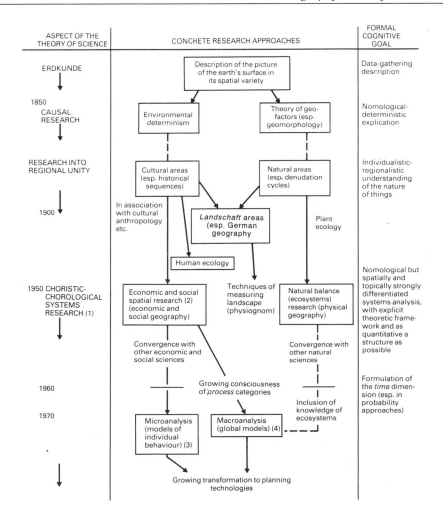

Figure 5.9   The development of research in geography
Source: From Bartels, 1973

sity teachers and for entry into certain professions (law, medicine, etc.). They also supplied qualified schoolteachers for disciplines in schools corresponding to those at university. It was hence primarily in countries where geography became an important school subject that it achieved academic status and the respect of the general public.

Since the Second World War, traditional university research (constrained as it is within the framework of its separate disciplines) has been increasingly supplemented by problem-orientated or **applied research**. Applied research often reflects the organization of government funding bodies rather than the traditional university subject. A new and interdisciplinary job market has been created at the same time as the number of traditional, discipline-orientated posts in both universities and schools is declining. In planning, for example, posts are advertised at both national and local levels that do not require training in any one specific discipline at the same time as the number of staff employed in many university departments is being reduced. Consequently, the discipline-orientated structure of the universities has come under severe stress. The essentially democratic decision-making processes of European universities have, to some extent, fended off these developments, but the hierarchical structure of, particularly, American universities has led to the closure of many single-subject departments and has encouraged the establishment of sector-orientated centres such as in regional and environmental studies. To an increasing extent, the European universities – challenged by sectorally orientated education in other higher education institutions (the former polytechnics in the UK, district high schools in Scandinavia) – are lowering the barriers between their disciplines and establishing new, integrated patterns of study.

Granö (*ibid.*) argues that these developments should suit geography very well, since geography has less easily adapted itself to discipline-orientated research than to the problem-orientated research the learned geographical societies always supported. To the extent that geography has maintained its cosmographic structure within university institutions, it may be able to play an effective role in modern problem-orientated contract research, provided that geographers are outward-looking, active and ready to co-operate with colleagues who can offer specialisms different from their own.

The tendency of physical geography to become more involved in applied research is particularly promising in this context. Orme (1985, p. 264) points out that **physical geography** (while lacking any distinct status in so far as its raw materials are shared by other natural sciences) can still play an important role through its emphasis on synthesis. The task is to shape a working synthesis of data (from geomorphological, climatological and biogeographical studies) into a comprehensive expression of the physical environment. However, it would not be enough to stop there: 'The physical geographer must also be practical and keen to apply findings to such areas as resource management, regional development and urban planning' (*ibid.*, p. 265). Exclusive specialist research on, for example the morphological effects of the last ice age on the Scottish landscape would be less acceptable in such a programme, and narrowly focused doctoral research has declining relevance in the present job market. It seems clear, however, that geography departments are well suited for work in the broad field of applied research and could profit from a working **geographical synthesis**.

Weichhart (1975) attempted to present a solution for the problem of geographical synthesis (see Figure 1.8, p. 14) which provides useful guidelines for the sort of applied research projects that are of growing importance in geography. Many German geography departments use the threefold division of complex geography as a framework for those degree courses which train planners for public and private service.

EXPLANATION AND DESCRIPTION

Adherents of the **spatial science** school criticized geography for being short on theory and long on facts, but in Chapter 3 we maintained that the spatial science school itself might be criticized for developing greater refinement of description rather than explanations: *the aim of geographical research is to describe and explain the world we perceive.* This world consists of **unique** but not **singular** places: each is formed through a unique interpretation of and adaptation to general structures (Box 5.3, pp. 155–57). Spatial scientists of a positivist tendency were content to study surface appearances and they refrained from structural analysis, which many social scientists have regarded as the real way to explanation. Increasingly, however, critics have focused on the fact that many spatial and structural models are unrealistic in that they ignore the complexity of real-world situations, concentrating on idealized postulates as rational economic behaviour. It is important to focus on the fact that 'individuals have different attributes and motivations and respond in varying and very often non-economic ways to different environmental characteristics' (Walmsley and Lewis, 1993, p. 5). Walmsley and Lewis (*ibid.*, p. 6), emphasizing the **behavioural** aspect of human geography, maintain that a better understanding of humanity–environment interaction may be achieved by studying the processes through which individuals come to know the environment in which they live, and how these processes influence resultant behaviour. At the microscale, **humanistic approaches** may also serve to offer a route to explanation.

At the macroscale, **structuralism** has been suggested as a route to explanation. Gray (1975, p. 231) claims that the structures identified by Marxist analysis (such as capitalist economic development and its relationships) are 'the underlying and missing variable causing the surface of reality which geographers examine'. But as pointed out by Johnston (1980), the process of capitalist development is such a general trend with so many independent variables that it cannot explain any particular 'real' spatial pattern: 'Thus, if one accepts the existence of the process one must accept any pattern as a realisation of that process: the existence of the process is not falsifiable' (*ibid.*, p. 408). In conclusion, it might be suggested that general theories – either spatial or historical – do not offer ultimate solutions to the problems of geographical explanation. The one ignores the underlying processes of change; the other neglects visible geographical variations. Geography must explain not only general processes and spatial structures but also particular geographical realizations.

It could even be maintained that, in a frenetic search for grand explanations, we have forgotten the value of the descriptions that enlighten us. During the last decades, geographers have been so strongly socialized into the theoretical social sciences that they have lost many of their former contacts with other idiographic sciences, such as history, biology and geology. Description is a legitimate scientific endeavour as long as it creates new knowledge.

ARE DICHOTOMIES OVEREMPHASIZED?

In conclusion, it would be foolish to exaggerate any dichotomy within the discipline. Positivist as well as critical approaches are, for instance, both of value. The majority of geographers belong neither to one camp nor the other, for both sets of ideas concern us. The changing balance between geographers represents variations in the emphasis placed on different elements of meta-theory. As for 'scientific revolutions', in the sense that Kuhn described them, in geography they have been more like changes in intonation. The basic elements of the old **paradigms** have not been rejected but carried along in some other form. Geography and similar disciplines have not totally rejected old theories in the same way as may have happened in theoretical physics (Kuhn's original specialism).

Dichotomies such as the one between **positivism** and **critical theory** fascinate us because they provide us with simplistic illusions of having understood something significant. When it comes to the crunch, however, truth is not simple. James (1972, pp. 506–7) considered that accepting the many dichotomies is a semantic trap, and that the attribution of fixed meanings or interpretations to word symbols may result in unreal conflicts. He suggested that the following dichotomies, in particular, have damaged geographical thinking. That

1) geography must be either idiographic or nomothetic, but not both;
2) physical and human geography are clearly differentiated branches of the discipline with separate concepts and methods;
3) geography must be either systematic or regional;
4) geographical methods must be either inductive or deductive;
5) geography must be classified as either a science or an art (the positivist–humanist discussion); and we might add that
6) we must rely on either a quantitative or a qualitative methodology.

In fact, geography straddles all these dichotomies.

Geography, like many other disciplines, must be both idiographic and nomothetic. There is a continuous transfusion from one to the other: most research contains something of both. An integration of physical and human geography is one of the tasks that justifies the existence of the subject. The division between systematic and regional geography is only expedient on strictly practical grounds; one merges into the other. Both induction and deduction (or merged in abduction) should, in general, be used in the same scientific analysis. The research problem and not any dogmatic view, should guide the choice of qualitative and/or quantitative methods.

'LET ONE HUNDRED FLOWERS BLOOM'

Hermeneutic methods may be invaluable in the right context in just the same way as **empirical-analytical methods** derived from the 'geometrical sciences' are appropriate to other investigations. The models and theories developed in diffusion and location studies, for example, provide important measuring rods for an improved understanding of geographical reality. When studying contemporary situations that are subject to change, we should also realize that the results of our research may influence the development of society. Such cases involve the problems of **double hermeneutics**: what we reveal is only what we have understood or modelled. Those directly concerned with the situation under review, however, may act upon a very limited and crude understanding of our explanations. From their crude understanding they may change the situation, hence making new scientific explanations possible. This process of change and explanation might continue in a never-ending hermeneutic spiral. As research workers, geographers belong to the society that is being studied and are thus part of its general understandings, beliefs and ways of life. A **critical** approach, involving a search for the structures that enable and constrain, is thus of crucial importance.

Each of these approaches and kinds of knowledge is of relevance to geography: the empirical-analytic derived from the natural sciences; the historical-hermeneutic derived from arts and history; the double hermeneutics derived from critical theory as explained by Jürgen Habermas and his colleagues; the **structuration theories** as explained by Giddens; and **realism** as developed by Sayer and others. The extent to which we rely on each of them should be related to the nature of our research problem rather than to our own strong dogmatic prejudices. **Poststructuralism** may help here to prevent us from assuming a dogmatic position.

We should approach the problem as to whether the history of geography has been influenced by idealistic or materialistic considerations (see p. 97) in much the same way. It will be clear from the comments above that, in the author's view, materialist considerations and also the nature of contemporary society (or 'spirit of the age' – *Zeitgeist*) influence science to a very large extent. It is easiest to demonstrate these influences, however, during the earliest phases of scientific development: from antiquity through the Middle Ages to the nineteenth century. The breakthrough of the positivist ideal, however, changed this: science achieved a relative independence that made it much easier for individual scholars to influence research ideas and learning. The contributions of Vidal de la Blache in France, of Schlüter in Germany, of Hägerstrand in Sweden, of Haggett and Sayer in Britain cannot be regarded as having been conditioned by the material base. The work of Marxist-inspired geographers living in capitalist countries demonstrates the limited constraints provided by the structural mechanisms of the economic base. The relative freedom of science has never been greater, and nor have the opportunities for the individual scientist to contribute to the development of his or her own discipline. However, other conditions (including the quantity of academic publications and the degree of specialization within science) make it difficult for the researcher to reach the entire potential audience. Like all other citizens, however, scientists live in a society that will try to control the scientists' activities to some extent, especially in so far as taxpayers generally supply the necessary research funds. In North America especially, where funds are to a great extent derived from business sources, the research worker may experience a conflict of loyalties if he or she feels that the big firm (or the government agency) has interests incompatible with those of the population as a whole. So while the individual scientist eventually becomes involved in politics just like any other citizen, the development of science itself is conditioned by both the abilities of individual research workers and by changes in society at large.

James (1972) argues that much of the antagonism between geographers is more of a battle of words than of realities. Ever since the subject's birth each new generation of geographers has asserted that it has founded a new geography, each overstressing the differences between the 'new' and the 'old'. This is a reason for asking, coldly and plainly, 'What *is* really new?' A study of the discipline's development shows that it may not have experienced any fundamental paradigm shifts. There have been changes in intonation – for example, in the shift of interest towards nomothetics or towards innovations in techniques, methods or conceptual thinking. A change sometimes occurs when a whole new world of information makes it necessary for us to alter our perceptions, but 'new' concepts are seldom as wholly new as their discoverers would lead us to believe – although they can contribute useful additions to our knowledge. The argument that new knowledge builds up and extends the old is in direct opposition to Kuhn's idea of paradigm shifts. When a new generation maintains there has been a scientific revolution, this is often the result of overenthusiasm for the quality of the material they have produced and also because the proponents of the 'revolution' have forgotten – or have chosen not to notice – what the old tradition stood for.

As noted by Taylor (1976), when we look carefully at any 'revolution' in geography, it will display some of the characteristics that are typical of scientific 'revolutions' in general. The new generation will use the strategy of making its contributions seem so difficult to understand that it is unlikely the 'old guard' will be able to master them. It is evident that some within the critical tradition disregarded those who were not familiar with the critical tradition's methods of argument and concepts of philosophy, much in the same way as Haggett's *Locational Analysis in Human Geography* (1965) deliberately discouraged those who were not familiar with mathematics and statistics.

We should not avoid real logical and philosophical difficulties, but neither should we express ourselves in unnecessarily complicated language or symbols so that our message only reaches a part of our potential audience. One of the traditional ideals we have in common with history and other 'arts' subjects is the need to express ourselves as clearly and as simply as possible. Geography is an academic discipline with academic obligations, but it is also a discipline (as is history) with obligations to inform the general public. This means the discipline has both to fulfil public expectations as well as to advance academic research.

SPECIALIZATION AND PLURALISM

The broad field of inquiry traditionally attributed to geography requires that research workers in the subject deploy a wide variety of skills; it also requires that research workers recruited to the discipline have experience in mathematics, statistics, biology and geology, as well as in history, sociology and economics. As no one individual can hope to cover more than a couple of these fields, it has been necessary to build up a staff of specialists in each of the branches of geography so that the whole discipline can be presented to students. The adage that 'geographers specialize in not being specialists' in no way applies to the staff of university geography departments. Research workers must specialize in order to create something worthwhile; in the normal course of events, an individual will work in a field that interests him or her and in which he or she has a fair amount of background knowledge. As an individual's specialization develops, he or she will resent the imposition of any kind of paradigm – even if that paradigm states only that the aim of geography should be as a discipline of **synthesis**.

James Bird (1979, p. 118) observes that there 'certainly are basic strains within geography, and if one paradigm is plastered across the subject, it will soon be broken by the disjunctions below'. It is, he believes, a hopeful sign at present that the idea of a ruling paradigm (which is similar in effect to an imposed orthodoxy) has been more or less discarded. Alternative schools of thought coexist, and this diversity is a good thing because it offers understanding with wider dimensions.

Gerhard Hard (1973, p. 237) suggests that, with our increasing awareness of the multitude of scientific traditions that are pursued within the framework of geography, we will begin to doubt the extent to which there has been a single former geographic discipline. Neither a consecutive, cumulative story nor a dialectical approach which identifies paradigms and revolutions will provide a wholly satisfactory account of what has happened. Our perspective on a discipline's history is always more or less influenced by the norms and outlook of the present generation. We see history in the terms of the present. Whether we emphasize the continuity and gradual growth of a science or dwell on its discontinuities and revolutions, we tacitly assume a single line of progression to our present situation; perhaps we should stress geography's heterogeneity – its many-faceted and rich traditions.

Hard (*ibid.*) took a close look at Fenneman's circles (Figure 1.1): are real geographers only those who integrate all the branches of the discipline into their research? If so, there are very few real geographers. Hard uses Venn diagrams to develop his argument (Figure 5.10). His first example, though simple, is not totally unrealistic: the term 'real geographer' may here only be applied to those who are committed to a) geomorphology or b) cultural landscape morphology, or both (shaded in Figure 5.10 (I)). If, however, synthesis is the sole aim of geography, a 'real geographer' must study both the physical and the cultural morphology of the landscape to qualify (Figure 5.10 (II)). If we consider the research themes actually pursued by self-styled geographers, the figure expands (Figure 5.10 (III)) to include climatologists, geomorphologists, biogeographers, ecogeographers and landscape geographers of different sorts, as well as economic geographers, locational theorists, behavioural geographers and so on. It is too restrictive (Figure 5.10 (IV)) to include among the 'real geographers' only those ecogeographers who are attempting a synthesis of natural and human factors.

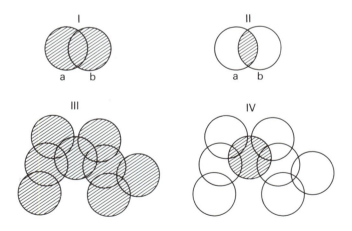

Figure 5.10   Different definitions of a 'geographer' (shaded). The circles symbolize closely connected research themes: (a) geomorphologists; (b) landscape morphologists; I-IV are explained in the text
Source: Adapted from Hard, 1973, p.235

Box 5.11   The cultural landscape – functional, intentional and structural modes of explanation

Figure 5.11   The relationship between the factors that contribute to an explanation of the cultural landscape
Source: Jones (1988)

Geography is a field-study discipline that has a keen interest in the relations between humanity and nature as these may be studied in the everyday cultural landscapes that surround us. One focus of investigation (promoted by, among others, Carl Sauer – see Box 2.9) has been the study of changes in the cultural landscape, in land tenure, land use, building culture and settlement form (i.e. very much a focus on physical changes and their explanation).

However, Jones (1988, p. 203) points out that in such a study we need three modes of explanation: functional, structural and intentional (see Figure 5.11)

> to explain the form and patterns of the cultural landscape. This can be represented as a triangle in which the three sides represent the three modes of explanation and the associated geographical scales (micro, meso and macro) . . . At the corners are phenomena which appear to provide a link between them: the individual actor as the link between intention and function; production as the link between function and structure; and ideology as the link between structure and intention.

As an example of this we might consider a study of a farming landscape, with all its in-fields and outfields, forests and farm buildings. The *functional* mode of explanation focuses on the landscape's functions in human terms. This is more or less a traditional geographical perspective in which modes of production are related to natural resources, buildings, property structures, market possibilities, etc. Such studies tend to draw their inspiration from positivism and the natural sciences and they work mostly at the meso-scale; the cultural landscape of a region.

In the *intentional* mode of explanation, a particular feature of a cultural landscape (such as land use) can be explained 'in a chronological-biographical sense in relation to the needs, motives, preferences, decisions and actions of individual persons' (*ibid.*, p. 200). A building may be explained through its history and through the needs and

Figure 5.12 A traditional Norwegian farm

aspirations of the farmer. Here we see the farmer as 'agency', acting on the basis of personal intentions and his or her interpretations of the functional resources at hand as well as the structural forces at the macroscale (e.g. the threats of globalization and of changing national policies).

Whereas the intentional mode of explanation focuses at the microlevel on the individual agent, *structural* explanations try to explain changing landscape features as adjustments to agricultural policies, technological innovations or market forces. Hence explanation is sought here at the macrolevel in relation to socioeconomic structures and related ideologies.

We need all three modes of explanation to understand such a basic and simple geographical object of study as the cultural landscape. But we must, of course, accept that the individual researcher may focus his or her research project on only one of these modes. Aasbø (1997) points out that there is currently an increasing interest within geographical research in the intentional mode. There is tendency to push this perspective further towards the study of identity and meaning in the **cultural landscape. Insiders** (e.g. farmers using the landscape) and **outsiders** (e.g. environmental planners and tourists) have conflicting perspectives on the identity and meaning contained within a cultural landscape which might lead to misunderstandings and antagonistic policies.

One reason for this new focus in the research seems to be the economic changes that have transformed our society in the last hundred years. In his time, Vidal de la Blache could study the cultural landscape dominated by primary industries, the features of the cultural landscape could be understood locally as results of the **vertical con-nections** between humanity and the natural conditions in that landscape. Hence a **chorological** perspective based on the notion of **absolute space** was not problematic.

The growth of manufacturing industry meant the breakdown of local dependencies, making a focus on **relative space** more fruitful. The cultural landscape could now be understood better in terms of transport cost – as a result of the physical distances between factors of production, of the location of raw materials and markets, and of the availability and location of labour. Geography became a 'discipline in distance'.

The 'third industrial revolution' (which is, rather, the advent of a **postindustrial society** in which only a fraction of the population really produces goods) has created new and less directly explicable landscape features. Figure 5.12 shows a traditional Norwegian farm surrounded by its in-fields and forests. Looking at the picture you might overlook the people when trying to work out why the farm buildings are built as they are, consider the farm's production potential and its market possibilities. However, the picture attracts our attention for a quite different reason. There are a lot of people sitting on the sloping field watching a scene where young people are performing traditional folk dances. And in front of the barn there is a van containing a TV crew. Even though it is raining slightly (as witnessed by the umbrellas), people have paid money to be here. This is a **media-dominated landscape**. The farmer may receive income from this activity, but the farm and the farm buildings function only as a stage set!

This brave new world of ours is continuously throwing up new tasks and problems for the geographer to solve. Humanity and nature are still dependent on each other but local connections are no longer so easy to see. The global dependencies between humanity and the fragile environment of the earth become, however, more and more clear. We have to think globally, as stated in the WCED report *Our Common Future*, but the necessary local actions are often obscured by the diffuse effects of these dependencies on the global, and even local, environment.

Capelle (1979, p. 64) suggested that Fenneman's geography was an inward-looking discipline, whereas 'geography today has become outward-looking, although not aggressively so, towards compatible sister sciences on its periphery'. Hartshorne (1939, pp. 243–5) described geography as an 'integrative' discipline, but that is not the same as regarding it as an 'integrated' discipline. There is an integrative task to transmit impulses from branch to branch within the structure, but there is no need to withdraw from the periphery.

We may still postulate a core or nerve centre, but to regard regional geography as this core is no longer helpful. The core in spatial science terms was defined by Ackerman (1963, p. 433) as 'thinking geographically:

> To structure his mind in terms of spatial distributions and their correlations is a most important tool for anyone following our discipline. The more the better. If there is any really meaningful distinction among scientists, it is this mental structuring. It is one reason why we should approach analogues from other fields, as from physics, with the utmost care. The mental substrates for inspiration differ from field to field.

This mental structuring is still our trademark, but today we should like to proceed below the surface appearance of spatial distributions and analyse processes and actions that lead to events in space and place. Our concern will be to analyse processes and how they are linked to both structural necessities and local, place-bound factors. We will be interested in how our **relational space** is formed in our everyday life. (Jones's (1988) approach to the study of the **cultural landscape** as presented in Box 5.11, p. may help to illustrate this further.) I believe that an integrative effort to keep all the fast-emerging branches on our periphery within the mother science should be our common purpose. Although at the research frontier an integrated synthesis may be an impossible task, in practical life, in local planning for example, the situation is rather different. Here the need for certain types of geographical synthesis is apparent. Synthesis is also a realistic aim where simplification rather than refinement is the objective. For this reason, basic teaching in geography at schools and in universities should cover both physical and human aspects of the subject.

I have (Holt-Jensen, 1988), with Weichhart (1975), stressed **ecogeography** as integrative geography, not because it should be regarded as geography proper but because it is an important field at the present time when humanity's use of the natural resources of the world may well determine humanity's future. My personal view is that the discipline of geography can make an essential contribution to the planning of a future for humankind, and that in this context geography should base its analysis of humanity's activity more on ecological than on economic analogies. The reader, however may prefer an alternative view, and a number are offered.

Perhaps none of the schools of thought in contemporary geography will emerge as 'the geography of the future', but we should try to retain them all within the structure of the discipline. As to the future, we may add, following Bird (1979, p. 124), that 'the geographer will never be the master of his fate: his very reason always progresses by leading him into the unknown and unforeseen where he learns new things'.

References and Bibliography

Aasbø, S. (1997) *Landskap på Bortsiao. Landskapsidentitet og modernisering i et ruralt utkantområde*. MA thesis, Department of Geography, University of Bergen.

Aase, A. (1970) Geografi og samfunn. Noen tendenser og problemer i dagens samfunnsgeografi, *Norsk Geografisk Tidsskrift*, Vol. 24, pp. 1–21.

Aase, T. H. (1999) The theological construction of conflict. Gilgit, northern Pakistan, in L. Manger (ed.) *Local Islam*, Curzon Press, London.

Aase, T. H. (1999) The use of metaphor in Himalayan resource management, in S. Toft Madsen (ed.) *State, Society and the Environment in South Asia*, Curzon Press, Honolulu, HI.

Abler, P., Adams, J. S. and Gould, P. (1972) *Spatial Organization: The Geographer's View of the World*, Prentice-Hall, London.

Ackerman, E. A. (1945) Geographic training, wartime research and immediate professional objectives, *Annals of the Association of American Geographers*, Vol. 35, pp. 121–43.

Ackerman, E. A. (1958) *Geography as a Fundamental Research Discipline*, University of Chicago, Department of Geography, research paper no. 53.

Ackerman, E. A. (1963) Where is a research frontier? *Annals of the Association of American Geographers*, Vol. 53, pp. 429–40.

Adams, J. D (1968) *A Review of Behaviour and Location*, LSE Graduate School discussion paper no. 20, London.

Agnew, J. A. (1987) *Place and Politics: The Geographical Mediation of State and Society*, Allen & Unwin, Boston, MA.

Agnew, J. A. and Duncan, J. S. (1989) *The Power of Place*, Unwin Hyman, Boston, MA.

Allen, J. and Massey, D. (1995) *Geographical Worlds*, The Open University, Oxford University Press.

Alvesson, M. and Sköldberg, K. (1994) *Tolkning och reflektion. Vetenskapsfilosofi och kvalitativ metod*, Studentlitteratur, Lund.

Andrews, H. F. (1986) The early life of Paul Vidal de Blache and the makings of modern geography, *Transactions, Institute of British Geographers* NS, Vol. 11, pp. 174–82.

Appleton, J. (1975) *The Experience of Landscape*, Wiley, Chichester.

Åquist, A. C. (1981) *Kuhns paradigmteori. Ett försök till tillämpning på kulturgeografi*, Rapporter og notiser 61, Institutt för Kulturgeografi, Lund.

Asheim, B. T. (1987) A critical evaluation of postwar developments in human geography in Scandinavia, *Progress in Human Geography*, Vol. 11, pp. 333–53.

Asheim, B. T. (1990a) How to confuse rather than guide students: a review of Holt-Jensen's *Geography – History and Concepts, Progress in Human Geography*, Vol. 14, pp. 281–92.

Asheim, B. T. (1990b) Norsk kultur – samfunnsgeografisk forskning i et nordisk perspektiv, in E. Barkved and A. Holt-Jensen (eds.) *Geografiens stilling i det post-industrielle samfunn – undervisnings- og forskningsoppgaver*, Bergen, Skrifter fra Norske Geografers Forening, Ny serie nr. 1.

Axelsen, B. and Jones, M. (1987) Are all maps mental maps? *GeoJournal*, Vol. 14, pp. 447–64.

Bagrow, L. (1945) The origin of Ptolemy's 'Geographia', *Geografiska Annaler* pp. 318–87.

Banse, E. (1924) *Die Seele der Geographie Geschichte einer Entwicklung*, Braunschweig und Hamburg.

Barrows, H. H. (1923) Geography as human ecology, *Annals of the Association of American Geographers*, Vol. 13, pp. 1–14.

Bartels, D. (1968) *Zur wissenschaftstheoretischen Grundlegung einer Geographie des Menschen*, Beihefte zur geographische Zeitschrift 19 F, Steiner, Wiesbaden.

Bartels, D. (1973) Between theory and metatheory, in R. J. Chorley (ed.) *Directions in Geography*, Methuen, London.

Batterbee, R. W., Flower, R. J., Stevenson, J. and Rippey, B. (1985) Lake acidification in Galloway: a palaeological test of competing hypotheses, *Nature*, Vol. 314, pp. 350–2.

Baudrillard, J. (1987) *Forget Foucault*, Semiotexte, New York.

Beck, H. (1982) *Grosse Geographen. Pioniere–Aussenseiter–Gelehrte*, Dietrich Reimer Verlag, Berlin.

Bennett, R. G. (1996) Challenges in Norwegian coastal zone planning, *GeoJournal*, Vol. 39, pp. 153–65.

Bennett, R. J. (1985) Quantification and relevance, in R. J. Johnston (ed.) *The Future of Geography*, Methuen, London.

Bennett, R. J. and Chorley, R. J. (eds.) (1978) *Environmental Systems: Philosophy, Analysis and Control*, Methuen, London.

Bennett, R. J. and Wrigley, N. (eds.) (1981) *Quantitative Geography: A British View*, Routledge & Kegan Paul, London.

Beresford, M. (1951) The lost villages of medieval England, *Geographical Journal*, Vol. 117, pp. 129–47.

Berry, B.J.L. (1973a) A paradigm for modern geography, in R.J. Chorley (ed) *Directions in Geography*, Methuen, London.

Berry, B.J.L. (1973b) *The Human Consequences of Urbanization*, Macmillan, London.

Berry, B.J.L. (1974) Review of D. Harvey *Social Justice and the City*, *Antipode*, Vol. 6, pp. 142–5, 148.

Bhaskar, R. (1975/1978) *A Realist Theory of Science*, Leeds Books, Leeds (reprinted 1978 by Harvester, Brighton).

Bhaskar, R. (1979) *The Possibility of Naturalism*, Harvester Press, Hassocks.

Biilmann, O. (1981) *Geografi, tradisjoner og perspektiver*, Geografforlaget, Brenderup.

Bird, J. (1975) Methodological implications for geography from the philosophy of K. R. Popper, *Scottish Geographical Magazine*, Vol. 91, pp. 153–63.

Bird, J. (1977) Methodology and philosophy, *Progress in Human Geography*, Vol. 1, pp. 104–10.

Bird, J. (1979) Methodology and philosophy, progress report, *Progress in Human Geography*, Vol. 3, pp. 117–20.

Bird, J. (1993) *The Changing Worlds of Geography* (2nd edn), Oxford University Press, Oxford.

Bjørlykke, C. (1996) Islands and oil – locality and structure, *GeoJournal*, Vol. 39, pp. 221–7.

Bliss, L. W. *et al.* (1969) *Laboratory Manual for General Botany* (4th edn), Holt Rinehart & Winston, New York.

Boal, F. W. (1969) Territoriality and the Shankill–Falls divide, Belfast, *Irish Geography*, Vol. 6, pp. 30–50.

Bobek, H. and Schmithüsen, J. (1949) Die Landschaft im logischen System der Geographie, *Erdkunde*, Vol. 3, pp. 112–20.

Bordieu, P. (1977) *Outline for a Theory of Practice*, Cambridge University Press, Cambridge.

Bordieu, P. and Wacquant, L.J.D. (1992) *An Invitation to Reflexive Sociology*, Polity Press, Cambridge.

Braithwaite, R. B. (1953) *Scientific Explanation*, Cambridge University Press, Cambridge.

Breitbart, M. M. (1981) Peter Kropotkin, the anarchist geographer, in D. R. Stoddart (ed.) *Geography, Ideology and Social Concern*, Blackwell, Oxford.

Broek, J.O.M. (1965) *Geography, its Scope and Spirit*, Merrill, Colombus, OH.

Broek, J.O.M. and Webb, J. W. (1973) *A Geography of Mankind*, McGraw-Hill, New York.

Brunhes, J. (1912) *La geographie humaine, essai de classification positive, principes et exemples*, Paris.

Bunge, W. (1962) *Theoretical Geography* (2nd edn 1966), Lund Studies in Geography, Series C1, Gleerup, Lund.

Bunge, W. (1966) Annals commentary: locations are not unique, *Annals of the Association of American Geographers*, Vol. 56, pp. 375–6.

Bunge, W. (1973) Ethics and logic in geography, in R. J. Chorley (ed.) *Directions in Geography*, London, Methuen.

Burton, I. (1963) The quantitative revolution and theoretical geography, *The Canadian Geographer*, Vol. 7, pp. 151–62.

Buttimer, A. (1978) Charisma and context: the challenge of la Géographie Humaine, in D. Ley and M. S. Samuels (eds.) *Humanistic Geography, Prospects and Problems*, Maaroufa Press, Chicago, IL.

Buttimer, A. (1981) On people, paradigms and 'progress' in geography, in D. R. Stoddart (ed.) *Geography, Ideology and Social Concern*, Blackwell, Oxford.

Buttimer, A. (1983) *The Practice of Geography*, Longman, London and New York.

Buttimer, A. and Seamon, D. (eds.) (1980) *The Human Experience of Space and Place*, Croom Helm, Beckenham.

Capel, H. (1981) Institutionalization of geography and strategies of change, in D. R. Stoddart (ed.) *Geography, Ideology and Social Concern*, Blackwell, Oxford.

Capelle, R. B. Jr (1979) On the periphery of geography, *Journal of Geography*, Vol. 78, pp. 64–8.

Carlstein, T. *et al.* (1978) *Timing Space and Spacing Time* (3 vols), Arnold, London.

Chabot, G. (1950) Les conceptions françaises de la science géographique, *Norsk Geografisk Tidsskrift*, Vol. 12, pp. 309–21.

Chisholm, M. (1975) *Human Geography: Evolution or Revolution?* Pelican, Harmondsworth.

Chorley, R. J. (1973a) Geography as human ecology, in R. J. Chorley (ed.) *Directions in Geography*, Methuen, London.

Chorley, R. J. (1976) Some thoughts on the development of geography from 1965 to 1975, in D. Pepper and A. Jenkins (eds.) *Proceedings of the 1975 National Conference on Geography in Higher Education. Oxford Polytechnic Discussion Paper in Geography*, Oxford Polytechnic Press, Oxford.

Chorley, R. J. and Haggett, P. (eds.) (1965) *Frontiers in Geographical Teaching*, Methuen, London.

Chorley, R. J. and Haggett, P. (eds.) (1967) *Models in Geography*, Methuen, London.

Christaller, W. (1933) *Die zentralen Orte in Süddeutschland*, Jena (English trans. C. W. Baskin (1966) *Central Places in Southern Germany*, Prentice-Hall, Englewood Cliffs, NJ).

Christaller, W. (1968) Wie ich zur der Theorie der zentralen Orte gekommen bin, *Geographische Zeitschrift*, Vol. 56, pp. 88–101.

Claval, P. (1980) Epistemology and the history of geographical thought, *Progress in Human Geography*, Vol. 4, pp. 371–84.

Claval, P. (1996) From a 'cultural' world to a 'political' one, in I. Douglas, R. Huggett and M. Robinson (eds) *Companion Encyclopedia of Geography*, Routledge, London.

Cloke, P., Philo, C. and Sadler, D. (1991) *Approaching Human Geography. An Introduction to Contemporary Theoretical Debates*, Paul Chapman, London.

Coates, B. E., Johnston, R. J. and Knox, P. L. (1977) *Geography and Inequality*, Oxford University Press, Oxford.

Coates, B. E. and Rawstron, E. M. (1971) *Regional Variations in Britain*, Batsford, London.

Collingwood, R. (1946) *The Idea of History*, Oxford University Press, Oxford.

Cooke, P. (edit) (1989) *Localities: The Changing Face of Urban Britain*, Unwin Hyman, London.

Cornu, A. (1955) *Marx et Engels*, Presses Universitaires de France, Paris.

Cosgrove, D. (1984) *Social Formation and Symbolic Landscape*, Croom Helm, London.

Cosgrove, D. and Daniels, S. (1988) *The Iconography of Landscape*, Cambridge University Press, Cambridge.

Coull, J. R. (1996) Towards a sustainable economy for the Shetland Islands: development and management issues in fishing and fish farming, *GeoJournal*, Vol. 39, pp. 185–94.

Dangermound, J. (1992) What is a geographical information system? in Johnson, A.I., Petterson, C.B. and Fulton, J.L. (eds) *Geographic Information Systems and Mapping – Practices and Standards*, Philadelphia, American Society for Testing and Materials.

Darwin, C. (1859) *The Origin of Species*, London.

Davies, W. K. (1972) *The Conceptual Revolution in Geography*, University of London Press, London.

Dear, M. (1988) The postmodern challenge: reconstructing human geography, *Transactions, Institute of British Geographers*, Vol. 13, pp. 262–74.

Demangeon, A. (1905) *La Picardie et les régions voisines, Artois, Cambresis, Beauvaises*, Armand Colin, Paris.

Dicken, P. (1998) *Global Shift* (3rd edn), Paul Chapman, London.

Dickinson, R. E. (1939) Landscape and society, *Scottish Geographical Magazine*, Vol. 55, pp. 1–14.

Dickinson, R. E. (1969) *The Makers of Modern Geography*, Routledge & Kegan Paul, London.

Dickinson, R. E. (1970) *Regional Ecology, the Study of Man's Environment*, Wiley, New York.

Dorling, D. and Fairbairn, D. (1997) *Mapping: Ways of Representing the World*, Longman, Harlow.

Dray, W. H. (1966) *Laws and Explanation in History*, Oxford University Press, Oxford.

DuBois, W.E.B. (1899, republished 1967) *The Philadelphia Negro: A Social Study*, Benjamin Bloom, New York.

Dunbar, G. S. (1981) Elisée Réclus, an anarchist geographer, in D. R. Stoddart (ed.) *Geography, Ideology and Social Concern*, Blackwell, Oxford.

Duncan, J. S. (1985) Individual action and political power: a structuration perspective, in R. J. Johnston (ed.) *The Future of Geography*, Methuen, London.

Duncan, N. (ed.) (1996) *Body Space. Destabilising Geographies of Gender and Sexuality*, Routledge, London.

Duncan, S. S. and Goodwin, M. (1988) *Uneven Development and the Local State*, Polity Press, Cambridge.

Engels, F. (1969) *The Conditions of the Working Class in England*, Panther, St Albans.

Entrikin, N. J. (1976) Contemporary humanism in geography, *Annals of the Association of American Geographers*, Vol. 66, pp. 615–32.

Entrikin, N. (1991) *The Betweenness of Place*, Johns Hopkins University Press, Baltimore, MD.

Eyles, J. (1985) *Senses of Place*, Silverbrook Press, Warrington.

Fawcett, C. B. (1919) *The Provinces of England* (rev. edn 1960), Hutchinson, London.

Febvre, L. (1922) *La terre et l'evolution humaine*, in the series *L'Evolution de l'Humanité, Paris*, English trans. 1925, *A Geographical Introduction to History*, Knopf, London.

Fenneman, N. M. (1919) The circumference of geography, *Annals of the Association of American Geographers*, Vol. 9, pp. 3–11. Reprinted in F. E. Dohrs, L. M. Sommers and D. R. Petterson (eds.) (1958) *Outside Readings in Geography*, Crowell, New York.

Fischer, E., Campbell, R. D. and Miller, E. S. (1969) *A Question of Place: The Development of Geographic Thought*, Beatty, Arlington, VA.

Fleure, H. J. (1919) Human regions, *Scottish Geographical Magazine*, Vol. 35, pp. 94–105.

Fochler-Hauke, G. (ed.) (1959) *Geographie*, Das Fischer Lexikon, Frankfurt.

Forer, P. (1978) A place for plastic space, *Progress in Human Geography*, Vol. 2, pp. 230–67.

Fosso, E. J. (1997) Industristeders arbeidstilbud og generasjoners forhold til utdanning, arbeid og sted – eksempelet Årdal, PhD thesis, Department of Geography, University of Bergen.

Foucault, M. (1972) *The Archaeology of Knowledge*, Tavistock, London.

Foucault, M. (1980) *Power/Knowledge: Selected Interviews and other Writings, 1972–1977*, C. Gordon (ed.), Harvester, Brighton.

Foucault, M. (1986) Of other spaces. *Diacritics*, Vol. 16, pp. 22–7.

Freeman, T. W. (1961) *A Hundred Years of Geography*, Duckworth, London.

Freeman, T. W. (1980) The Royal Geographical Society and the development of geography, in E. H. Brown (ed.) *Geography, Yesterday and Tomorrow*, Oxford University Press, Oxford.

Freeman, T. W. and Pinchemel, P. (eds.) (1978) *Geographers, Bibliographical Studies, Vol. 2*, Mansell, London.

Freeman, T. W. and Pinchemel, P. (eds.) (1980) *Geographers, Bibliographical Studies, Vol. 4*, Mansell, London.

Friedmann, J. (1992) *Empowerment*, Blackwell, Oxford.

Garrison, W. L. (1959–60) Spatial structure of the economy, *Annals of the Association of American Geographers*, Vol. 49, pp. 232–9, 471–82; Vol. 50, pp. 357–73.

Gerland, G. (1887) Die wissenschaftliche Aufgabe der Geographie, ihre Methode und ihre Stellung im praktischen Leben, *Beiträge zur Geophysik*, Vol. 1, pp. 4–54.

Giddens, A. (1979) *Central Problems in Social Theory*, Macmillan, London.

Giddens, A. (1984) *The Construction of Society*, Polity Press, Oxford.

Gilbert, A. (1988) The new regional geography in English- and French-speaking countries, *Progress in Human Geography*, Vol. 12, pp. 208–28.

Goudie, A. (1990) *The Human Impact on the Natural Environment* (3rd edn), Blackwell, Oxford.

Gould, P. (1979) Geography 1957–1977: the Augean period, *Annals of the Association of American Geographers*, Vol. 69, pp. 139–51.

Gould, P. (1985) *The Geographer at Work*, Routledge & Kegan Paul, London.

Gould, P. and White, R. (1974) *Mental Maps*, Penguin, Harmondsworth.

Gourou, P. (1936) *Les Paysans du delta Tonkinois*, École Française d'Extrême-Orient, Paris.

Gradmann, R. (1931) *Süd-Deutschland* (2 vols), J. Engelhorn, Stuttgart.

Granö, J. G. (1929) Reine Geographie, *Acta Geographica*, Vol. 2, pp. 1–202.

Granö, O. (1981) External influence and internal change in the development of geography, in D. R. Stoddart (ed.) *Geography, Ideology and Social Change*, Blackwell, Oxford.

Granö, O. (1986) *Finnish Geography 1880–1980, Highlight of the Decades*, University of Turku (mimeo).

Granö, O. (1987) Vetenskapens institutionella struktur och geografins utvekling, *Nordisk Samhällsgeografisk Tidsskrift*, Vol. 5, pp. 3–8.

Gray, F. (1975) Non-explanation in urban geography, *Area*, Vol. 7, pp. 228–35.

Greene, T. M. (ed.) (1957) *Kant Selections*, Scribner, New York.

Gregory, D. (1978) *Ideology, Science and Human Geography*, Hutchinson, London.

Gregory, D. (1989) Areal differentiation and post-modern human geography, in D. Gregory and R. Walford (eds.) *Horizons in Human Geography*, Macmillan, London.

Gregory, D. (1994) *Geographical Imaginations*, Blackwell, Oxford.

Gregory, K. J. (1985) *The Nature of Physical Geography*, Edward Arnold, London.

Grigg, D. (1965) The logic of regional systems, *Annals of the Association of American Geographers*, Vol. 55, pp. 465–91.

Guelke, L. (1974) An idealist alternative in human geography, *Annals of the Association of American Geographers*, Vol. 64, pp. 193–202.

Guelke, L. (1977a) The role of laws in human geography, *Progress in Human Geography*, Vol. 1, pp. 376–86.

Guelke, L. (1977b) Regional geography, *The Professional Geographer*, Vol. 29, pp. 1–7.

Guelke, L. (1978) Geography and logical positivism, in D. T. Herbert and R. J. Johnston (eds.) *Geography and the Human Environment, Progress in Research and Applications. Vol. 1*, Wiley, Chichester.

Guelke, L. (1981) Idealism, in M. E. Harvey and B. P. Holly (eds.) *Themes in Geographic Thought*, Croom Helm, Beckenham.

Guelke, L. (1982) *Historical Understanding in Geography: An Idealistic Approach*, Cambridge University Press, Cambridge.

Guess, R. (1981) *The Idea of a Critical Theory: Habermas and the Frankfurt School*, Cambridge University Press, Cambridge.

Gutersohn, H. (1945) *Harmonie in der Landschaft. Schriftenfolge der Schweizerischen Vereinigung für Landesplanung.* Schrift Nr. 1. Vogt-Schild AG, Solothurn.

Habermas, J. (1978) *Knowledge and Human Interest* (2nd edn), Heinemann, London.

Hägerstrand, T. (1953) Innovations – förloppet ur korologisk synpunkt, *Medd. Från Lunds Universitets Geografiska Institution. Avhandling* nr. 25 (trans. A. Pred (1967) *Innovation Diffusion as a Spatial Process*, University of Chicago Press, Chicago, IL).

Haggett, P. (1965) *Locational Analysis in Human Geography*, Arnold, London.

Haggett, P. (1983) *Geography: A Modern Synthesis* (3rd edn), Harper & Row, New York.

Haggett, P. (1990) *The Geographer's Art*, Blackwell, Oxford.

Haggett, P. (1996) Geographical futures. Some personal speculations, in T. Douglas, R. Huggett and M. Robinson (eds.) *Companion Encyclopedia of Geography*, Routledge, London and New York.

Haggett, P., Cliff, A. D. and Frey, A. (1977) *Locational Analysis in Human Geography*, Arnold, London.

Haines-Young, R. (1989) Modelling geographical knowledge, in B. Macmillan (ed.) *Remodelling Geography*, Blackwell, Oxford.

Haines-Young, R. and Petch, J. (1986) *Physical Geography: Its Nature and Methods*, Paul Chapman, London.

Hannerberg, D. (1961; 1968) *Att Studera Kulturgeografi*, Scandinavian University Books, Stockholm.

Hansen, F. (1994) Analyseramme – fagopfattelse og geografi, in J. Öhman (ed.) *Traditioner i Nordisk kulturgeografi*, Nordisk Samhällsgeografisk Tidskrift, Uppsala.

Hard, G. (1973) *Die Geographie, eine wissenschaftstheoretische Einfürung*, DeGruyter, Berlin.

Harloe, M., Pickvance, C. G. and Urry, J. (eds.) (1990) *Place, Policy and Politics. Do Localities Matter?* Unwin Hyman, London.

Harris, C. D. and Ullman, E. L. (1945) The nature of cities, *Annals of the American Academy of Political and Social Sciences*, Vol. 242, pp. 7–17.

Hart, J. F. (1982) The highest form of the geographer's art, *Annals of the Association of American Geographers*, Vol. 72, pp. 1–29.

Hartshorne, R. (1939) *The nature of geography: a critical survey of current thought in the light of the past*, *Annals of the Association of American Geographers* Vol. 29, pp. 173–658.

Hartshorne, R. (1950) The functional approach in political geography, *Annals of the Association of American Geographers*, Vol. 40, pp. 95–130.

Hartshorne, R. (1955) 'Exceptionalism in Geography' re-examined, *Annals of the Association of American Geographers*, Vol. 45, pp. 205–44.

Hartshorne, R. (1959) *Perspective on the Nature of Geography*, Rand McNally, Chicago, IL.

Hartsock, N.C.M. (1983) The feminist standpoint: developing a ground for a specifically feminist historical materialism, in S. Harding and M. B. Hintikka (eds.) *Discovering Reality*, Reidel, Dordrecht and London.

Harvey, D. (1969) *Explanation in Geography*, Arnold, London.

Harvey, D. (1973) *Social Justice and the City*, Arnold, London.

Harvey, D. (1974) Population, resources and the ideology of science, *Economic Geography*, Vol. 50, pp. 256–77.

Harvey, D. (1984) On the history and present condition of geography: an historical materialist manifesto, *The Professional Geographer*, Vol. 36, pp. 1–11.

Harvey, D. (1985a) *The Urbanization of Capital*, Blackwell, Oxford.

Harvey, D. (1985b) *Consciousness and the Urban Experience*, Blackwell, Oxford.

Harvey, D. (1989) *The Condition of Postmodernity: An Enquiry into the Origins of Cultural Changes*, Blackwell, Oxford.

Harvey, M. E. and Holly, B. P. (eds) (1981) *Themes in Geographic Thought*, Croom Helm, Beckenham.

Hassinger, H. (1919) Über einige Aufgaben geographischer Forschung und Lehre, *Kartographische und Schulgeographische Zeitschrift*, Vol. 8, pp. 65–76.

Healey, P. (1998) Institutionalist theory, social exclusion and governance, in J. Allen, G. Cars and A. Madanipour (eds.) *Social Exclusion in European Cities*, Jessica Kingsley, London.

Hegel, G.W.F. (1975, new edition) *Lectures on the Philosophy of World History*, Cambridge University Press, Cambridge.

Henriksen, G. (1973) *Grunnlagsproblemer og interaksjon - en metageografisk analyse*, Hovedfagsoppgave i geografi, Geografisk Institutt, Universitetet i Bergen.

Herbert, D. T. and Raine, J. W. (1976) Defining communities within urban areas, *Town Planning Review*, Vol. 47, pp. 325–38.

Hettner, A. (1927) *Die Geographie, ihre Geschichte, ihr Wesen und ihre Methoden*, Ferdinand Hirt, Breslau.

Hettner, A. (1929) Methodische Zeit- und Streitfragen. Neue Folge, *Geographische Zeitschrift*, Vol. 35, pp. 264–86, 332–44.

Hettner, A. (1930) Zur 'Stellungnahme von Seiten der Schulgeographie', *Geographische Anzeiger*, Vol. 31, pp. 353–6.

Holt-Jensen, A. (1988) *Geography: History and Concepts* (2nd edn), Paul Chapman, London.

hooks, b. (1992) *Black Looks: Race and Representation*, Turnaround, London.

Howard, E. (1902) *Garden Cities of Tomorrow*, London.

Huggett, R. J. (1993) *Modelling the Human Impact in Nature. Systems Analysis of Environmental Problems*, Oxford University Press, Oxford.

Huggett, R. and Robinson, M. (1996) Introduction, in I. Douglas, R. Huggett and M. Robinson (eds.) *Companion Encyclopedia of Geography. The Environment and Humankind*, Routledge, London.

Huntington, E. (1915) *Civilization and Climate*, Yale University Press, New Haven, CT.

Huxley, T. H. (1877) *Physiography*, Macmillan, London.

Jackson, P. (1989) *Maps of Meaning: An Introduction to Cultural Geography*, Unwin Hyman, London (reprinted 1992 and 1994 by Routledge, London).

James, P. E. (1972) *All Possible Worlds: A History of Geographical Ideas*, Odyssey Press, Indianapolis, IN.

Johannessen, K. S. (1985) *Tradisjoner og skoler i moderne vitenskapsfilosofi*, Sigma, Bergen.

Johansson, I. (1973) Anglosaxisk vetenskapsfilosofi, in *Positivism, Marxism, kritisk teori*, Pan/Nordsteds, Stockholm.

Johnston, R. J. (1968) Choice in classification, the subjectivity of objective methods, *Annals of the Association of American Geographers*, Vol. 58, pp. 575–89.

Johnston, R. J. (1978) Paradigms and revolutions or evolution, *Progress in Human Geography*, Vol. 2, pp. 189–206.

Johnston, R. J. (1980) On the nature of explanation in human geography, *Transactions, Institute of British Geographers*, Vol. 5, pp. 402–12.

Johnston, R. J. (edit 1985) *The Future of Geography*, Methuen, London.

Johnston, R. J. (1986a) *Philosophy and Human Geography* (2nd edn), Arnold, London.

Johnston, R. J. (1986b) *On Human Geography*, Blackwell, Oxford.

Johnston, R. J. (1986c) Four fixations and the quest for unity in geography, *Transactions, Institute of British Geographers*, NS, Vol. 11, pp. 449–453.

Johnston, R. J. (1991) *A Question of Place. Exploring the Practice of Human Geography*, Blackwell, London.

Johnston, R. J. (1997) *Geography and Geographers. Anglo-American Human Geography since 1945* (5th edn), Arnold, London.

Jones, M. (1988) Land-tenure and landscape change in fishing communities on the outer coast of central Norway, *c.* 1880 to the present. *Geografiska Annaler*, Vol. 70B, pp. 197–204.

Kant, E. (1946) Den indre omflyttingen i Estland i samband med de estniska städernas omland, *Svensk Geografisk Årsbok*, Vol. 22, pp. 83–124.

Kant, E. (1951) Omlandsforskning och sektoranalys, in G. Enequist (ed.) *Tätorter och Omland*, Lunde-quistska Bokhandelen, Uppsala.

Kearney, M. (1991) Borders and boundaries of the state and self at the end of the empire. *Journal of Historical Sociology*, 4(1), pp. 52–74.

Keltie, J. S. (1886) Geographical education: report to the Council of the Royal Geographical Society, *Supplementary papers of the Royal Geographical Society*, No. 1, pp. 439–594.

Keltie, J. S. (1921) Obituary for Peter Alexievich Kropotkin, *Geographical Journal*, Vol. 57, pp. 316–19.

Knox, P. L. (1975) *Social Well-Being: A Spatial Perspective*, Clarendon Press, Oxford.

Kropotkin, P. (1885) What geography ought to be, *Nineteenth Century*, Vol. 18, pp. 940–56.

Kropotkin, P. (1924) *Ethics: Origin and Development*, New York.

Kuhn, T. S. (1962/1970a) *The Structure of Scientific Revolutions*, University of Chicago Press, Chicago, IL.

Kuhn, T. S. (1970b) Reflections on my critics, in I. Lakatos and A. Musgrave (eds.) *Criticism and the Growth of Knowledge*, Cambridge University Press, Cambridge.

Lange, G. (1961) Varenius über die Grundfrage der Geographie, *Petermanns Geographische Mitteilungen*, Vol. 105, pp. 274–83.

Lash, S. and Urry, J. (1987) *The End of Organised Capitalism*, Polity Press, Cambridge.

Leser, H. (1980) *Geographie, Das Geographische Seminar*, Westermann, Braunschweig.

Livingstone, D. N. (1992) *The Geographical Tradition*, Blackwell, Oxford.

Löfgren, A. (1996) *Om kvalitativ metod och fältarbete i geografi. Arbeider fra Geografisk Institutt, Universitet i Trondheim, New series C*, no. 4, Universitetet i Trondheim, Trondheim.

Lowenthal, D. (1985) *The Past is a Foreign Country*, Cambridge University Press, Cambridge.

Lowenthal, D. (1996) *Possessed by the Past. The Heritage Crusade and the Spoils of History*, Free Press, New York.

Lukermann, F. (1958) Towards a more geographic economic geography, *The Professional Geographer*, Vol. 10, pp. 1–10.

Lundberg, A. and Handegård, T. (1996) Changes in the spatial structure and function of coastal cultural landscapes, *GeoJournal*, Vol. 39, pp. 167–78.

Lynch, K. (1962) *Site Planning*, MIT Press, Cambridge, MA.

Mabogunje, A. (1976) Systems approach to a theory of rural-urban migration, in J. Beishon and C. Peters (eds.) *Systems Behaviour* (2nd edn), Harper & Row/Open University Press, London.

Mackinder, H. J. (1911) The teaching of geography from an imperial point of view, and the use which could and should be made of visual instruction, *Geographical Teacher*, Vol. 6, pp. 79–86.

Madanipour, A. (1998) Social exclusion and space, in J. Allen, G. Cars and A. Madanipour (eds.) *Social Exclusion in European Neighbourhoods*, Jessica Kingsley, London.

Mair, A. (1986) Thomas Kuhn: and understanding geography, *Progress in Human Geography*, Vol. 10, pp. 345–69.

Malmberg, T. (1980) *Human Territoriality*, Mouton, The Hague.

Marsh, G. P. (1864) *Man and Nature, or Physical Geography as Modified by Human Action*, Schribners, New York.

Martin, G. J. (1985) Paradigm change: a history of geography in the United States 1892–1925, *National Geographic Research*, Spring, pp. 217–35.

Massey, D. (1984) *Spatial Division of Labour: Social Structures and the Geography of Production*, Macmillan, London.

Massey, D. (1994) *Space, Place and Gender*, Polity Press, London.

Massey, D. and Allen, J. (eds.) (1984) *Geography Matters. A Reader*, Cambridge University Press, Cambridge.

Massey, D. and McDowell, L. (1994) A woman's place, in D. Massey *Space, Place and Gender*, Polity Press, London.

May, J. A. (1970) *Kant's Concept of Geography and its Relation to Recent Geographical Thought*, University of Toronto Press, Toronto.

Mead, W. (1954) Ridge and furrow in Buckinghamshire, *Geographical Journal*, Vol. 120, pp. 34–42.

Meesenburg, H. (1996) Man's role in changing the coastal landscapes in Denmark, *GeoJournal*, Vol. 39, pp. 143–52.

Meijer, H. (1981) *Zuyder Zee–Lake Ijssel*, Haag: IDG, Utrecht.

Minshull, R. (1970) *The Changing Nature of Geography*, Hutchinson, London.

Morrill, R. L. (1965) *Migration and the growth of urban settlement, Lund Studies in Geography*, Series B, 24, Gleerup, Lund.

Morrill, R. L. (1974) Review of D. Harvey, *Social Justice and the City*, *Annals of the Association of American Geographers*, Vol. 64, pp. 457–7.

Morrill, R. L. (1984) Recollection of the 'quantitative revolution's' early years: the University of Washington 1955–65, in M. Billinge, D. Gregory and R. Martin (eds.) *Recollection of a Revolution*, Macmillan, London.

Morrill, R. L. and Wohlenberg, E. H. (1971) *The Geography of Poverty in the United States*, McGraw-Hill, New York.

Myrdal, G. (1953) The relation between social theory and social policy, *British Journal of Sociology*, Vol. 23, pp. 210–42.

National Curriculum Geography Working Group (1989) *Interim Report*, Brighton, Department of Education and Science and the Welsh Office.

Neef, E. (1982) Geographie-einmal anders gesehen, *Geographische Zeitschrift*, Vol. 70, pp. 241–60.

Newcomb, R. M. (1979) *Planning the Past. Studies in Historical Geography*, Dawson/Archon, London.

Norberg-Schultz, K. (1984) *Genius Loci. Towards a Phenomenology of Architecture*, Rizzoli, New York.

Nordgård, A. (1972) *Korologiske metoder*, Geografisk Institutt, Oslo.

Olsson, G. (1974) Servitude and inequality in spatial planning: ideology and methodology in conflict, *Antipode*, Vol. 6, pp. 16–21 (reprinted in R. Peet (ed.) (1977) *Radical Geography*, Methuen, London).

Olsson, G. (1975) *Birds in Egg*, Michigan Geography Publication no. 15, Ann Arbor, MI.

Olsson, G. (1978) Of ambiguity or far cries from a memoralising mamafesta, in D. Ley and M. Samuels (eds.) *Humanistic Geography: Prospect and Problems*, Croom Helm, Beckenham.

Olwig, K. (1984) *Nature's Ideological Landscape: A Literary and Geographic Perspective on its Development and Preservation on Denmark's Jutland Heath*, Allen & Unwin, London.

Olwig, K. (1996) Recovering the substantive nature of landscape, *Annals of the Association of American Geographers*, Vol. 86, pp. 630–53.

Openshaw, S. (1991) A view of the GIS crisis in geography: or, using GIS to put Humpty-Dumpty back together again, *Environment and Planning A*, Vol. 23, pp. 621–8.

Orme, A. (1985) Understanding and predicting the physical world, in R. J. Johnston (ed.) *The Future of Geography*, Methuen, London.

Paasi, A. (1991) Deconstructing regions: notes on the scales of spatial life, *Environment and Planning A*, Vol. 23, pp. 239–56.

Palm, R. (1986) Coming home, *Annals of the Association of American Geographers*, Vol. 76, pp. 469–79.

Paffen, K. H. (1955) Die natürlichen Landschaften und Ihre räumliche Gliederung, *Forschungen. z dt. Landeskunde 68*.

Paterson, J.H. (1974) Writing regional geography, in C. Board *et al.* (eds.) *Progress in Geography*, Edward Arnold, London.

Peet, R. (ed.) (1977) *Radical Geography*, Methuen, London.

Peet, R. (1981) Spatial dialectics and Marxist geography, *Progress in Human Geography*, Vol. 5, pp. 105–10.

Peet, R. (1985) The social origins of environmental determinism, *Annals of the Association of American Geographers*, Vol. 75, pp. 309–33.

Peet, R. (1996) Structural themes in geographic discourse, in I. Douglas, R. Huggett and M. Robinson (eds.) *Companion Encyclopedia of Geography*, Routledge, London.

Peet, R. (1998) *Modern Geographical Thought*, Blackwell, Oxford.

Peet, R. and Watts, M. (eds.) (1996) *Liberation Ecologies: Environment, Development, Social Movements*. Routledge, London.

Penck, A. (1894) *Morphologie der Erdoberflache*, Stuttgart.

Penck, A. (1901–9) *Die Alpen in Eisalter, Vols 1–3*, Stuttgart.

Penck, A. (1928) Neuere Geographie, in *Zeitschrift der Gesellschaft für Erdkunde zu Berlin*, Berlin, pp. 31–56.

Peschel, O. (1870) *Neue Probleme der vergleichenden Erdkunde als Versuch einer Morphologie der Erdoberfläche*, Duncker & Humblot, Leipzig.

Pickles, J. (1985) *Phenomenology, Science and Geography*, Cambridge University Press, Cambridge.

Popper, K. (1970) Normal science and its dangers, in F. Lakotos and A. Musgrave (eds.) *Criticism and the Growth of Knowledge*, Cambridge University Press, Cambridge.

Popper, K. R. (1972) *Objective Knowledge*, Oxford University Press, Oxford.

Porteous, J.D. (1977) *Environment and Behaviour; Planning and Everyday Urban Lift*, Addison-Wesley, Reading, Mass.

Poster, M. (1984) *Foucault, Marxism and History: Mode of Production versus Mode of Information*, Polity Press, Cambridge.

Poulton, M. (1995) Affordable housing at an affordable (social) price, in G. Fallis *et al.* (eds.) *Home Remedies: Rethinking Canadian Housing Policy*, C. D. Howe Institute, Toronto.

Pred, A. (1967) *Behaviour and Location. Part I. Foundations for a Geographic and Dynamic Location Theory*, Gleerup, Lund.

Pred, A. (1969) *Behaviour and Location. Part II. Foundations for a Geographic and Dynamic Location Theory*, Gleerup, Lund.

Pred, A. (1984) Place as a historically contingent process: structuration and the time-geography of becoming places, *Annals of the Association of American Geographers*, Vol. 74, pp. 279–97.

Quaini, M. (1982) *Geography and Marxism*, Blackwell, Oxford.

Ratzel, F. (1882) *Anthropogeographie. I. Oder Grundzüge der Anwendung der Erdkunde auf die Geschichte*, Engelhorn, Stuttgart.

Ratzel, F. (1891) *Anthropogeographie. II. Die geographische Verbrietung des Menschen*, Engelhorn, Stuttgart.

Ratzel, F. (1897) *Politische Geographie*, Oldenburg, Munich.

Réclus, E. (1866–7) *La Terre*, Hachette, Paris.

Réclus, E. (1875–94) *Nouvelle Géographie Universelle*, Hachette, Paris.

Réclus, E. (1905–8) *L'Homme et la Terre* (6 vols), Paris.

Reenberg, A. (1982) *Det katastroferamte Sahel*, Geografforlaget, Brenderup.

Relph, E. (1976) *Place and Placelessness*, Pion, London.

Relph, E. (1996) Place, in I. Douglas, R. Huggett and M. Robinson (eds.) *Companion Encyclopedia of Geography*, Routledge, London.

Ritter, C. (1822–59) *Die Erdkunde, im Verhältnis zur Natur und zur Geschichte des Menschen, oder allgemeine vergleichende Geographie als sichere Grundlage des Studiums und Unterrichts in Physikalischen und historischen Wissenschaften* (19 vols), Reimer, Berlin.

Sack, R. D. (1972) Geography, geometry and explanation, *Annals of the Association of American Geographers*, Vol. 62, pp. 61–78.

Sack, R. D. (1974) Chorology and spatial analysis, *Annals of the Association of American Geographers*, Vol. 64, pp. 439–52.

Sack, R. D. (1980) *Conceptions of Space in Social Thought*, Macmillan, London.

Sack, R. D. (1986) *Human Territoriality, Its Theory and History*, Cambridge University Press, Cambridge.

Sack, R. D. (1992) *Place, Modernity, and the Consumer's World*, John Hopkins University Press, Baltimore, MD.

Sauer, C. O. (1925) The morphology of landscape, *University of California Publications in Geography*, Vol. 2, pp. 19–35.

Sauer, C. O. (1963) *Land and Life* (ed. J. B. Leighley), University of California Press, Berkeley, CA.

Sayer, A. (1984, 1992) *Method in Social Science. A Realist Approach*, Hutchinson, London.

Sayer, A. (1985) Realism and geography, in R. J. Johnston (ed.) *The Future of Geography*, Methuen, London.

Sayer, A. (1991) Behind the locality debate: deconstructing geography's dualisms, *Environment and Planning A*, Vol. 23, pp. 283–308.

Sayer, A. (1993) Postmodernist thought in geography: a realist view. *Antipode*, Vol. 25, pp. 320–40.

Schaefer, F. (1953) Exceptionalism in geography, *Annals of the Association of American Geographers*, Vol. 43, pp. 226–49.

Schlüter, O. (1906) *Die Ziele der Geographie des Menschen*, München, Berlin.

Schlüter, O. (1920) Die Erdkunde im Verhältnis zu der Natur- und Geisteswissenschaften, *Geographische Anzeiger*, Vol. 21, pp. 145–52, 212–21.

Schmieder, O. (1964) Alexander von Humboldt: Persönlichkeit, wissenschaftliches Werk und Auswirkung auf die moderne Länderkunde, *Geographische Zeitschrift*, Vol. 52, pp. 81–95.

Schmithüsen, J. (1976) *Allgemeine Geosynergetik*, De Gruyter, Berlin.

Scholten, A. (1980). Al-Muqaddasi *c.* 945–*c.* 988, in T. W. Freeman and P. Pinchemel (eds.) *Geographers, Bibliographical Studies, Vol. 4*, Mansell, London.

Schültz, H. D. (1980) Die deutschsprachige Geographie von 1800 bis 1970, *Abhandlung des geographischen Instituts-Antropogeographie* Band 29, Selbstverlag des geographischen Instituts der Freien Universität, Berlin.

Schumacher, E. F. (1974) *Small Is Beautiful: A Study of Economics as if People Mattered*, Abacus, London.

Schutz, A. (1982) *Life Forms and Meaning of Structures*, Routledge & Kegan Paul, London.

Semple, E. C. (1911) *Influences of Geographical Environment*, Henry Holt, New York.

Shurmer-Smith, P. and Hannam, K. (1994) *Worlds of Desire, Realms of Power. A Cultural Geography*, Arnold, London.

Sibley, D. (1995) *Geographies of Exclusion*, Routledge, London.

Simpson, G. G. (1963) Historical science, in C. C. Albritton (ed.) *The Fabric of Geology*, Addison Wesley, Reading, MA.

Skjervheim, H. (1974) Objectivism and the study of man, *Inquiry*, Vol. 17, pp. 213–39, 265–302.

Smith, A. (1776) *An Inquiry into the Nature and Causes of the Wealth of Nations*, London.

Smith, D. M. (1979) *Where the Grass is Greener: Living in an Unequal World*, Penguin, Harmondsworth.

Smith, N. (1990) *Uneven Development: Nature, Capital and the Reproduction of Space*, Blackwell, London.

Soja, E. (1980) The socio-spatial dialectic, *Annals of the Association of American Geographers*, Vol. 70, pp. 207–25.

Soja, E. (1989) *Postmodern Geographies: The Reassertation of Space in Social Theory*, Verso, London.

Somerville, M. (1848) *Physical Geography*, London.

Sømme, A. (ed.) (1965) *Fjellbygd og feriefjell*, Cappelen, Oslo.

Speak, S., Gilroy, R., Cameron, S. and Woods, R. (1995) *Young Single Mothers: Barriers to Independent Living*, Family Policy Studies Centre, London.

Stam, R. (1992) Mobilizing fictions: the Gulf war, the media and the recruitment of the spectator, *Public Culture*, Vol. 4, pp. 101–26.

Stamp, D. (1966) Ten years on, *Transactions, Institute of British Geographers*, Vol. 40, pp. 11–20.

Steers, J. A. (1946) *The Coastline of England and Wales*, Cambridge University Press, Cambridge.

Steiner, D. (1965) Die Faktorenanalyse: eine modernes statistisches Hilfsmittel des Geographen für die objektive Raumgliederung und Typenbildung, *Geographica Helvetica*, Vol. 20, pp. 20–34.

Stevenson, W. I. (1978) Patrick Geddes 1854–1932, in T. W. Freeman and P. Pinchemel (eds.) *Geographers, Bibliographical Studies, Vol. 2*, Mansell, London.

Stewart, J. Q. (1947) Empirical mathematical rules concerning the distribution and equilibrium of population, *Geographical Review*, Vol. 37, pp. 461–85.

Stoddart, D. R. (1966) Darwin's impact on geography, *Annals of the Association of American Geographers*, Vol. 56, pp. 683–98.

Stoddart, D. R. (1967) Organism and ecosystem as geographical models, in R. J. Chorley and P. Haggett (eds.) *Models in Geography*, Methuen, London.

Stoddart, D. R. (1975) 'That Victorian science' – Huxley's 'physiography' and its impact on geography, *Transactions, Institute of British Geographers*, Vol. 66, pp. 17–40.

Stoddart, D. R. (1981) The paradigm concept and the history of geography, in D. R. Stoddart (ed.) *Geography, Ideology and Social Concern*, Blackwell, Oxford.

Stoddart, D. R. (1986) *On Geography*, Blackwell, Oxford.

Stoddart, D. R. (1987/1996) To claim the high ground: geography for the end of the century, *Transactions, Institute of British Geographers* NS, Vol. 12, pp. 327–36 (reprinted in S. Daniels and R. Lee (eds.) *Exploring Human Geography. A Reader*, Edward Arnold, London).

Taaffe, E. J. (ed.) (1970) *Geography*, Prentice-Hall, Englewood Cliffs, NJ.

Tatham, G. (1951) Geography in the nineteenth century, in G. Taylor (ed.) *Geography in the Twentieth Century*, Methuen, London.

Taylor, G. (ed.) (1951) *Geography in the Twentieth Century*, Methuen, London.

Taylor, P. J. (1976) An interpretation of the quantification debate in British geography, *Transactions, Institute of British Geographers* NS, Vol. 1, pp. 129–42.

Taylor, P. J. (1985) The value of a geographical perspective, in R. J. Johnston (ed.) *The Future of Geography*, Methuen, London.

Thrift, N. J. (1983) On the determination of social action in space and time, *Environment and Planning D: Society and Space*, Vol. 1, pp. 23–57.

Thrift, N. J. (1986) The geography of international economic disorder, in R. J. Johnston and P. J. Taylor (eds.) *A World in Crisis. Geographical Perspectives*, Blackwell, Oxford.

Troll, C. (1947) Die geographische Wissenschaft in Deutschland in dem Jahren 1933 bis 1945, *Erdkunde*, Vol. 1, pp. 3–48.

Tuan Y.-F. (1971) Geography, phenomenology, and the study of human nature, *The Canadian Geographer*, Vol. 15, pp. 181–92.

Tuan, Y.- F. (1974, reprinted 1996) Space and place: humanistic perspective, in J. Agnew, D. N. Livingstone and A. Rogers (eds.) *Human Geography. An Essential Anthology*, Blackwell, Oxford.

Tuan Y.-F. (1976) Humanistic geography, *Annals of the Association of American Geographers*, Vol. 66, pp. 266–76.

Tuan Y.-F. (1977) *Space and Place: The Perspectives of Experience*, University of Minnesota Press/Arnold, London.

Tuan Y.-F. (1978) Literature and geography: implications for geographical research, in D. Ley and M. S. Samuels (eds.) *Humanistic Geography: Prospects and Problems*, Maaroufa, Chicago, IL.

Tuan Y.-F. (1980) *Landscapes of Fear*, Pantheon, New York; Blackwell, Oxford.

Uhlig, H. (1967) *Methodische Begriffe der Geographie, besonders der Landschaftskunde. Separat-Vorabdruck aus Westermanns Lexikon der Geographie*, Westermann, Braunschweig.

Uhlig, H. (1971) Organization and system of geography, *Geoforum*, Vol. 7, pp. 7–38.

Uhlig, H. (1973) Landschaftökologie, in *Das Wesen der Landschaft*, Wissenschaftliche Buchgesellschaft, Darmstadt, pp. 268–85.

Ullman, E. (1941) A theory of location for cities, *American Journal of Sociology*, Vol. 46, pp. 835–64.

Unwin, T. (1992) *The Place of Geography*, Longman, Harlow.

van Valkenburg, S. (1952) The German school of geography, in G. Taylor (ed.) *Geography in the Twentieth Century*, Methuen, London.

Varenius, B. (1650) *Geographia Generalis*, Amsterdam.

Vidal de la Blache, P. (1903) *Tableau de la géographie de la France*, Hachette, Paris.

Vidal de la Blache, P. (1917) *La France de l'Est*, Armand Colin, Paris.

Vidal de la Blache, P. (1921) *Principes de la Géographie humaine* (trans. 1926 as *Principles of Human Geography*, Constable, London).

von Humboldt, A. (1845–62) Kosmos: Entwurf einer physischen Weltbeschreibung (5 vols), Cotta, Stuttgart (English trans. E. C. Ohé (1849–58) H. G. Bohn, London).

von Thünen, J. H. (1826) *Der Isolierte Staat in Beziehung auf Landwirtschaft und Nationalökonomie*, Hamburg (English trans. C. M. Wartenburg (1966) *Von Thünen's Isolated State*, ed. P. Hall, Pergamon, Oxford).

Wagner, D. L. and Mikesell, (1962) *Readings in Cultural Geography*, University of Chicago Press, Chicago.

Waibel, L. (1933) Was verstehen wir unter Landschaftskunde? *Geographische Anzeiger*, Vol. 34, pp. 197–207.

Walmsley, D. J. and Lewis, G. J. (1993) *People and Environment: Behavioural Approaches in Human Geography* (2nd edn), Longman, Harlow.

Walter, E. (1988) *Placeways*, University of North Carolina Press, Chapel Hill, NC.

Wärneryd, O. (1977) Kulturgeografi-Samhällsgeografi, in *Samhällsventenskapliga Studiernas Historik*, Lund, pp. 20–31.

Warntz, W. (1959) *Towards a Geography of Price: A Study in Geo-Econometrics*, University of Pennsylvania Press, Philadelphia, PA.

Warntz, W. (1964) A new map of the surface of population potentials for the United States, 1960, *Geographical Review*, Vol. 54, pp. 170–84.

Weber, A. (1909) *Über der Standort der Industrien*, Tübingen (trans. C. Friederich (1929) *Alfred Weber's Theory of the Location of Industries*, Chicago University Press, Chicago, IL).

Weichhart, P. (1975) Gesucht: Eine human-ökologisch orientierte Teildisziplin der komplexen Geographie, *Berichte zur deutschen Landeskunde*, Vol. 54, pp. 125–32.

Weichhart, P. (1980) *Geographie im Umbruch*, Franz Deuticke, Wien.

Werlen, B. (1993) *Society, Action and Space. An Alternative Human Geography*, Routledge, London and New York.

White, G. (1973) Natural hazards research, in R. J. Chorley (ed.) *Directions in Geography*, Methuen, London.

White, G. (1974) Edward Ackerman 1911–73, *Annals of the Association of American Geographers*, Vol. 64, pp. 197–309.

White, R. and Gould, P. (1974) *Mental Maps*, Penguin, Harmondsworth.

Whittlesey, D. (1929) Sequent occupance, *Annals of the Association of American Geographers*, Vol. 19, pp. 162–5.

Widberg, J. (1978) *Geografi, från naturvetenskap til samhällsvetenskap: En idéhistorisk översikt* (3-betygsuppsats), Institutt för Kulturgeografi och Ekonomisk Geografi, Lund.

Wolpert, J. (1964) The decision process in spatial context, *Annals of the Association of American Geographers*, Vol. 54, pp. 337–58.

Wood, J., Gilroy, R., Healey, P. and Speak, S. (1995) *Changing the Way We Do Things here: The Cruddas Park Initiative*, Department of Town and Country Planning, Newcastle University, Newcastle upon Tyne.

World Commission on Environment and Development (WCED) (1987) *Our Common Future*, Oxford University Press, Oxford.

Wrigley, E. A. (1965) Changes in the philosophy of geography, in R. K. Chorley and P. Haggett (eds.) *Frontiers in Geographical Teaching*, Methuen, London.

Yeates, M. (1968) *An Introduction to Quantitative Analysis in Economic Geography*, McGraw-Hill, New York.

Author and Personality Index

Glossary and subject index

abduction   a scientific method (combining elements of *induction* and *deduction*) that is often applied in case study-based social research. The empirical study of a particular case may lead to surprising results. We try to formulate a tentative theory or *hypotheses* that, if correct, will explain the case-study results. However, such a tentative theory or hypotheses must be tested, corroborated or refuted through further case studies. The most crucial phase in abduction is the 'lift' from empirical results to tentative theory (*see also* Figure 3.5)   62, 67

abiotic   pertaining to or characterized by the absence of life or living organisms   12, 13

absolute space   an understanding of space as a distinct, physical and imminently real or empirical entity. Traditional *regional geography* studies the empirical entities, dependencies or vertical connections between humanity and the environment within the 'container space' of a particular region   73–74, 84, 146, 195

abstract theory   the isolation conceptually of a *partial* aspect of an object, or a conception that represents an ideal view of reality. In *spatial science* abstraction is the starting-point of model building. In *realism*, however, it is the identification of the *essential* characteristics inherent in the constitution of structures and the construction of the ideal types of relationships between structures, mechanisms and events which, ultimately, has to be tested through concrete research   130, 131

academic (in the sense of the definition of a discipline)   a discipline defined purely on the basis of what the 'avantgarde' of researchers find rewarding and interesting (often in contrast to what a *vernacular* definition implies)   139

action theory   a term employed by Werlen (1993) to urge geographers to focus their inquiry on actions rather than space. Geographers should focus their attention on actions and processes in space and how these are related to the agents' understanding of the subjective, social and physical worlds   147, 148, 161

agency   *see* human agency   125, 126, 132, 150, 151, 173

Allgemeine Geographie   a term used by German geographers to denote systematic geography, in some instances defined as the study of a single geofactor in a regional or global context   14

Allgemeine Geosynergetik   synergistic – working together, being co-operative. The term is used by Schmithüsen (1976) as an underlying philosophical premise in regional geography where human and natural factors work together in complex unities (regions)   4

Annales School   a school of history that aimed to study the history of ordinary people rather than the history of governments, kings, wars, etc.   48

Anthropogeographie   a German term denoting human geography   42, 43

applied geography   the application of current geographical knowledge and analytical techniques to solving contemporary problems in society   54, 116, 186

a priori model   a model that exists in the mind prior to being tested scientifically   64

axiom   a universally accepted principle or rule; an elementary statement; a basic building block in scientific work   102

behavioural geography   a form of geography that regards the knowledge and perceptions of decision-makers as equally or more important than the physical and economic conditions of the environment. *Cognition* is considered to be the key in understanding human spatial behaviour   93, 96, 147, 188

Berkeley school   a school of geography closely associated with Carl Sauer when he was chairman of the Geography Department at the University of California, Berkeley. Although the school admitted to no doctrinaire positions with regard to methods or subject-matter, it was associated with *cultural geography*, the *diffusion* of cultural traits and *cultural landscape* and the role of *human agency* in its creation. This school had a marked predeliction for *fieldwork*   51

biogeography   the systematic geographical study of plants and animals

biotic   pertaining to life   12, 13

case study   a social investigation – for example, of a single tract within an urban area selected *not* to provide statistical representativeness but rather to provide a typical (or revealing) account of the local pattern of everyday life   131

causal explanation   all observed facts have causes and, hence, science should look for causes not purposes. Science should set up hypotheses for possible causes that can be tested in a rigid scientific procedure; the hypothetic-deductive method   69, 70

central place theory   based on the work of Walter Christaller. The development of systems of central places within a settled area, the distribution of consequent tributary areas and the hierarchical relations between them   77, 78, 92, 93

chorography (*choros* = place; *graphos* = to draw, describe)   the description of places and regions; of things occurring in the same place   19, 21, 22, 24, 45, 68

chorology (*choros* = place; *logos* = rational structure, explanation)   the scientific study of the relationships between phenomena that occur in the same place   22, 24, 45, 50, 68, 83, 146, 195

chronology   the study of events that follow each other in time   24

cognitive description   a description of what is known or understood; rational ordering; the classification and analysis of data that have been collected   88

cognized environment   the environment as perceived and consciously understood   121

comparative method   the systematization and comparison of data, region by region   28

complex geography   the study of relationships between several geofactors in a regional or global context. Although akin to regional geography, complex geography does not necessarily analyse the full range of geofactors. It hence stands in contrast to *Allgemeine Geography*, which concentrates on the analyses of single geofactors   14

concrete research   empirical study of events and contingent conditions in *realism*   130, 131

contingency   a contingent event; a chance, accident or possibility that is conditional of something uncertain. In a simplified form of *realism*, a contingency is often understood as special (or accidental) local conditions that promote deviations from general trends   9, 130, 155

contingent condition   *see* contingency   130

corroborate   make more certain, confirm   66, 67

cosmography (*cosmos* = the world or universe conceived as an orderly system; *graphos* = to draw, describe)   a term used in the seventeenth and eighteenth centuries for the scientific description of the universe. Cosmography encompasses the sciences of cartography, geography, biology, geophysics and also anthropology   2, 21, 22, 26, 29, 40, 69, 185

crisis phase   a term employed by Thomas Kuhn to describe a phase in the development of scientific theory when new thinking forces a reappraisal of the significance of objective data and a vigorous debate among the scientists concerned   60

critical rationalism   the philosophy of science as developed by Karl Popper. Critical rationalism is based on the principle of falsification – a process of error elimination, critical testing and the possible corroboration of preconceived theories   65, 66, 100

critical theory   associated with the Frankfurt School and the later contributions of Jürgen Habermas. Critical theory criticizes *positivism*. Its central concern is the *historicity of social action*, particularly how human agency is influenced by modern capitalism. Critical theory goes beyond Marx's critique of political economy as it also includes elements of the hermeneutic tradition   98, 99, 100, 108, 109, 112, 189

cultural geography   a subfield of human geography that focuses on the material and non-material aspects of human culture (e.g. the study of the agricultural practice of growing paddy rice and the cultural traits of the religious beliefs that underlie such practice). Cultural geography was particularly associated with the *Berkeley School*. Today, it has been revitalized through 'new cultural geography', which focuses on the cultures of everyday life and the relationships between dominant and subordinate cultures   52, 93

cultural landscape   the landscape as influenced through time by the technology and attitudes of the human cultural groups who have inhabited it   6, 46, 55, 114, 193–196

cultural trend (in human geography)   *see* cultural geography   169, 173

cycle of erosion   the regular changes of landforms through time. Landscapes are uplifted as a result of forces within the earth and the movement of continents. They are later eroded by water, ice and wind until further uplifts renew the cycle. Uplift and erosion can take place simultaneously   41, 42

Darwinism   the application of the theories of evolution proposed by Charles Darwin that involve the development from simpler to more complicated lifeforms, the struggle between different species of plants and animals, natural selection and the random character of the many variations observed in nature   37, 69

dasein (being here)   awareness of the conditions of one's existence   150

deductive reasoning   the derivation of a particular truth from a more general statement. Starting from a general statement (e.g. the *hypothesis* that 'all swans are white'), the truth of this statement is tested by empirical observation. The opposite of *induction*   63, 64, 67

deep structures   the structures that underlie observed patterns   122

determinism   (see environmental determinism)   27, 68, 70, 113

deterministic model   a model that will produce only one particular result when a set of inputs is applied to that model   81

dialectic   a form of critical examination that is to be found in Plato and that was developed by Immanuel Kant to show the mutually contradictory character of scientific principles when these are applied to determine objects beyond the limits of experience. Georg Hegel applied the dialectic to the processes of thought through which such contradictions (theses and antitheses) eventually appear to be resolved within a higher truth   97, 105

diffusion   the spread of domesticated animals and crops, ideas and techniques over space and through time. The concept is particularly associated with studies undertaken by the Berkeley School (Carl Sauer) and the Lund School of Geography in Sweden (Torsten Hägerstrand)   81, 85

disciplinary matrix   an interpretation of the concept of a *paradigm* whereby a particular scientific community at a particular period of time shares an understanding of its discipline's tasks, special character and methods, which delimit it from other disciplines   59, 72, 185

discursive consciousness   the type of knowledge we can explain and discuss in words; connected to Giddens' structuration theory   126

double hermeneutics   the process by which people who are the subject of an attitudinal survey may change their views after having considered how the results of the study will be interpreted. A process of ongoing interaction and interpretation   109, 189

duality of structure   structures set the conditions for human action, but humans are skilled agents who can mould the institutions and other structures of society through their conscious and unconscious actions. Connected to Giddens' structuration theory   125, 126

ecogeography   the study of the systems of *geofactors* that arise as a result of the interaction between people and their environments   15, 88, 93, 176, 177, 196

ecological analysis   the study of the connections between environmental variables as well as between human and environmental variables   11, 12, 50

economic base   industries or other forms of production upon which complex economies are thought to be dependent   106, 123, 158

economic man   a model of human behaviour that assumes that the decision-maker has a perfect knowledge of cost factors and an ability to use that information to maximize returns and minimize costs. The use of this model in *location theory* made it possible for analysts to predict outcomes without reference to human behaviour. The use of the model in *spatial science* produced *normative models* of a highly mechanical nature   93, 94, 96

ecosystem   a mutually dependent system of plants, animals and organic and inorganic processes that exists in a specific area or place   88, 177, 178, 179

empirical-analytical science   a science (particularly a natural science, such as physics) that can be substantiated through the use of a precise scientific language and that can engender predictive knowledge for the technical control of society   108, 111, 112, 113, 117, 189

empirical generalization   a generalization based on facts concerning the here and now and, as such, that is only valid for a particular time and place. In positivistic terms, an empirical generalization is more scientific than factual statements or systematized descriptions   63, 64, 73

empirical question   a question about factual content; on what *is* rather than what *ought* to be   102

empiricism   the belief that our sense experiences provide us with the only true basis for knowledge   100, 102, 113, 115, 127

Enlightenment movement   a philosophical movement of the seventeenth and eighteenth centuries that is particularly associated with Voltaire, Rousseau and other French thinkers. It was characterized by its belief in the power of human reason and of universal education   100, 101

environmental determinism   the belief that human activities and cultures are profoundly influenced and constrained by the natural environment   38, 42, 43, 44

episteme   according to Michel Foucault, a regime of true or a 'general politics of truth' that is defined and accepted within a society in order to reinforce the power relations within that society   62, 107

epistemology   the application and development of knowledge; a theory of knowledge that seeks to determine correspondence between the realm of knowledge (concepts, propositions) and the realm of objects (experiences, things). An epistemology guides the formulation of research problems   115, 117, 118, 125, 128, 132, 134, 141

error elimination   in Popper's *critical rationalism*, the empirical testing of tentative theories to determine which theories have weak explanatory value   65, 66

everyday-life approach   a bottom-up research approach that lets people speak for themselves in order to elucidate the meanings, metaphors and ideas these people follow in their day-to-day lives   137, 150, 158, 174

evolution   the gradual transformation of organisms from lower and simpler to higher and more complicated forms   37, 38

eugenics   the 'science' of improving the qualities of the human race by the careful selection of 'suitable' parents   38

exemplar   a *paradigm* in the form of an *exemplary model*   58, 59, 60, 72, 85

exemplary model   a particular piece of scientific work that is used as a model for subsequent research   72, 85

existentialism   an individual's essential 'being in the world' and the irreducible personal dimensions of human life (*see* Box 5.1)   121, 150

factor analysis   a quantitative method of transforming a data matrix so that the variables in the new matrix are uncorrelated. The relative importance of each factor in explaining a certain phenomenon can hence be elucidated   82

falsification   the possibility of proving a scientific theory to be false. Karl Popper would not accept a theory as scientific if it was not conceivably possible to falsify it   64, 65, 100

fieldwork   a term traditionally used in geography for the methods by which primary data are obtained through observation in the place of study. Fieldwork became less important in human geography when secondary data (from censuses, etc.) became the focus of quantitative analysis. However, it was revitalized in the *case studies* typical of contemporary *cultural geography*   20, 52, 54, 70

Fliese   a mosaic tile; a term applied to the smallest category of subregions that form the building blocks of landscape morphology (*see* Figure 2.2)   51, 53

formal logics   the use of logics in *logical positivism*, by which rules are set for scientific reasoning. Built on certain given and universally accepted elementary statements or *axioms*, further statements of knowledge may be derived through empirical investigations and/or logical steps of argumentation or chains of evidence   101

Frankfurt School   a school of thought that was founded in Frankfurt in 1923 and that was re-established in New York in 1934. It later came to be associated with Jürgen Habermas, among others (*see critical theory*),   99, 100, 109

functional region   a region defined by its function (e.g. a city and the hinterland that uses the city as its service centre)   54, 85, 154

Ganzeit   totality   27, 39

Garden city movement   town planning ideas particularly connected to Ebenezer Howard   36

Gemeinschaft   a much-used German term that denotes the close, local relations between people in traditional, rural societies and the strict social control of, and care for, all members of that community   169, 170, 173

gender   social sex, behaviour, attitudes and opportunities that depend on socially constructed views of femininity and masculinity   155–158, 160, 166

general geography   a term used by Varenius to denote systematic geography   14, 23

generalization trap   the failure to recognize unique features – local conditions – in geographical study   143

genius loci   a concept used by Norberg-Schultz for the 'spirit of place' – the *phenomenological* understanding of the physical meaning and characteristics of a place   158, 159, 160

genre de vie   a central concept in traditional French regional geography. The way of life characteristic of a human group who have organized themselves to exploit the resources of the land on which they live   46, 48

geofactor   a factor that influences geographical distributions   10, 11, 12, 13, 51

geographical law   see scientific law

geographical societies   national and local societies, mostly founded in the nineteenth century, that promote geographic knowledge, expeditions and national interests   2, 29

geographical synthesis   combination of factors working together to create particular geographical features (see Figure 1.9)   2–16, 185, 187

geomorphology   the study of landforms   41, 42, 70

geopolitics   political geography based on Nazi ideology and Darwinistic ideas of the survival of the fittest. Hence the strongest, fittest nations have the right to rule over other, less strong and less fit nations   43, 45

geosphere   the round earth, the surface of the globe   12

Gesellschaft   a much-used German term that describes role-directed relationships in modern, urban societies. People have relations with other people in the same workplace as themselves or in a club of mutually shared interests, but not with their neighbours   170, 173

GIS (geographical information systems)   computer programs that make it possible to store map and other geographical data in a digitalized form and that facilitate the analysis of these data   175, 177, 179, 180–185

glacial morphology   study of land-forms created by glaciers in present or former times   41

globalization   the spread of industrial economies and technologies and the establishment of multinational firms across the world. Globalization has also come to mean the spread of a universal culture fostered through the modern mass media, which has created 'placeless' features – both physically and socially   71, 146, 154, 167, 174, 176

gravity model   a mathematical model that is used to study patterns of interaction. The model assumes that the interaction of people and goods between centres is generally proportional to the product of their populations (or output of goods) and inversely proportional to the distance between them   80, 122

grounded theory   a theory of science that is based on a reaction to crude positivism. Grounded theory is based on pragmatism, social relevance and *idiographic, case-study-based* indepth qualitative research rather than on quantitative, representative data sets. Grounded theory supports, however, the positivistic ideal of scientific objectivity and the possibility of separating theory from empirical data   112

hermeneutic approach   an interpretative (rather than explanatory) approach to understanding (*verstehen*) and to clarifying meaning   94, 95, 104, 112, 189

historical-hermeneutic science   arts and sciences that, as the history of art and the analysis of poetry, have to base their approach on the interpretation of meanings. A hermeneutic approach rather than an approach based on measurements and data analysis   108, 117

historical materialism   a programme of historical research propounded by Karl Marx and Friederich Engels based on the productive powers of societies and their tendency to outgrow current systems of social relationships. When this occurs, contradictions between the organization of production and society can be expected to lead to periods of social revolution, such as the rise of capitalism and its predicted replacement by socialism   105, 106, 123, 138

holism (holistic)   a belief that a whole (for instance an organism or a region) is more than the sum of its parts and that it is the task of science to grasp this totality   27, 30

horizontal connections/linkages   the links and dependencies between places (particularly in terms of goods and money), as between a centre and its market hinterland   74, 75, 143, 146, 154, 179

human agency   in Giddens' structuration theory, the relations between intentions and actions, including the reflexive monitoring of action, the rationalization of action and the motivation for action.

Human agency refers more to the agent's capabilities for action and the action itself than to the agent's intentions   125, 126

human ecology the study of the spatial aspects of human relations, particularly humanity–environment relations; closely related to the functional approach in geography   50, 177–179

humanistic geography a term used by Yi-Fu Tuan geographies based on the methods of the humanities that strive to synthesize subjective and objective studies and that lay particular importance on human creativity   93, 95, 117, 188

hypothesis a statement established for the purpose of empirical testing   63, 64, 66

hypothetic-deductive method this method sets up hypotheses to be tested in order to see how far they can explain observed reality. Satisfactory explanations lead towards the formulation of scientific laws   39, 57, 63, 64, 65, 67, 102

idealism the view that the things we perceive to be real are in fact contained within our own minds. As a philosophical stance it is opposed to *materialism* and *realism*. As a concept, however, it has different meanings. For instance, when used to describe science it implies that science is led by the high, noble principles of individual scientists, not by the material interests and forces in society. Idealism can also mean the tendency to represent things in an ideal form (or as they might be or should be), rather than as they are   28, 94, 97, 98, 106, 121, 127

identity of place in geography, identity of place has come to mean place as experienced by human beings – their evaluation of physical setting, human activity and meaning   159, 160, 163

ideological base religion, culture of a society (see also *superstructure*)   123

Ideologiekritik critique of prevailing doctrines of societies, to support the enlightenment of people to their own true interests   109

idiographic an explanation or description of unique or particular phenomena or connections. The opposite of *nomothetic*. An idiographic science searches for explanations of the single instance, such as a historic event that is not thought to be the result of general trends or the laws of history   10, 56, 57, 72, 73, 83, 104, 113

induction reasoning in which a conclusion is based on observed empirical data. Induction often leads to unfounded generalization: all swans I have seen are white; therefore all swans are white. The opposite of *deductive reasoning*   39, 46, 62, 63, 67

innovation the introduction of a new phenomenon, for instance a new farming practice. Geographical studies of innovations have focused on their origins and spread (*diffusion*) through time and over space   81, 93

insidedness as opposed to the outsidedness in a place and often related to the study of the *identity of place*   160, 194

institutionalization (of a scientific discipline) the establishment of scientific departments in order to educate students to prepare them to take part in a profession   28, 34, 35

instrumentalism models and laws concerned with the technical control of the environment that manipulate rather than explain   115

intensive concrete research a *realist approach* to research, as recommended by Sayer, in which a limited number of indepth, empirical *case studies* are undertaken to achieve an understanding of the functions of *necessary* and *contingent relations*   131

interpretant the participants in the *semiotic triad*. The people who communicate personal interpretations and understandings in the form of *signs*, such as words   134

isomorphic different but seemingly having the same form. Humans seem to act in the same manner as some of the primitive laws of physics (*see* gravity model)   80

karst morphological features in limestone regions, including caves and subterranean rivers   41

laissez-faire the idea that governments should interfere as little as possible in economic affairs; 'the survival of the fittest' will secure economic growth and general welfare   38

Lamarckianism a thesis connected with the French naturalist, Jean Baptiste Lamarck. Lamarck maintained that hereditary changes were the result of organisms' efforts to adapt themselves to the changing conditions encountered in their environments. As a supplement to Darwinism, neo-Lamarckian thinking has had a profound influence on geography – for example, that human nature can be changed via transformations in, and better planning of, the physical and social environment. Hence the importance of *place analysis*   38, 44

Land country, territory 48

land use actual use of land areas for different agricultural uses, transport, housing, industries, recreation etc. 55

Land Use Survey registration on actual land use on maps 55

Länderkunde regional geography. The geographical study of particular areas that are differentiated one from the other through the synthesis of their individual physical and human features 10, 48, 49

länderkundliches Schema a method in regional geography devised by Hettner in which a region is defined via a definitive order of succession of the different physical and human geographic features that constitute that region 51, 72

landscape chronology the scientific reconstruction of former landscapes or attempts to explain landscape evolution 49

landscape ecology the study of the connections between biological communities and their environment in a particular section of a landscape 49

landscape geography the study of cultural and natural landscapes, their particular features and their evolution 11

landscape morphology the study of the visible phenomena on the earth's surface, both natural and artificial 50, 51, 52, 71, 72, 77

Landschaftskunde a term developed by Siegfried Passarge to describe the geography of landscapes as created through time by the interaction of human and physical forces. *Landschaftskunde* involves the study of descending hierarchies, from major natural regions (that owe their identities to climatic and physical features) to subregions that are largely differentiated as a result of human activity 10, 11, 49

life-world culturally defined spatial and temporal setting or horizon of *everyday life* 160

locale a concept used by Giddens to denote a meaningful physical setting for social interaction. A *relational space* may range from a room, a street corner or a town to a territory occupied by a nation-state 127, 149

localization in geographical terms, locating spatial phenomena on a map 7

location theory a theory that seeks to account for the location of economic activities 77, 79, 93, 96

logical classification in Immanuel Kant's categorization, the ordering of phenomena according to their nature: plants are different from stones, etc. 24

logical positivism an elaboration of positivism associated with the Vienna Circle. Unlike Comtean *positivism*, logical positivism admits that some statements can be verified without recourse to empirical experience. Encompassing the rules of *formal logics*, it is possible to distinguish between analytical statements (i.e. a priori propositions whose truth is guaranteed by their internal definitions, as in mathematics)', and synthetic statements (whose truth has to be established by *hypothesis* testing) 100, 101, 116

Marxism (see *historical materialism*) 105–107

Marxist geography (*see also* radical geography) a trend in geography based on *historical materialism*, concerned particularly with the ways in which the production of space, place and landscape is a reflection of specific social formations (capitalism, feudalism, socialism) 93, 97, 123, 124

materialistic viewpoint the material base governs the development of society as well as scientific knowledge (in opposition to idealism) 97, 98

mechanical explanation phenomena and observations are understood as outcomes of given causes. Can be considered as the opposite to *teleological explanation* 27, 97, 98

mental map the mentally stored images individuals have of spaces and places which they draw upon in their interpretations of spatial desirabilities and in the organization of their spatial routines. Behavioural geographic research has tried to show, for instance, how the mental maps of one particular city differ according to the city dwellers' different ethnic origins 96, 114

metaphor a word or phrase applied to an object or concept it does not literally denote in order to suggest a comparison with the other object or concept (as in 'mighty fortress is our God'). Metaphor is also used to describe the situation where a group of people try to give a new phenomenon meaning by relating it to or comparing it with something already well known to them 135

metaphysics the consideration of first principles, including the nature of being and knowing (*ontology* and *epistemology*). These first principles lie outside our sense perceptions or, rather, cannot be tested by *empirical* investigations. *Logical positivism*, therefore, regards metaphysical statements as unscientific 100, 101

method triangulation the evaluation of (and often the combination of) different methods to solve a defined research problem. Triangulation is used here as a *metaphor* – finding something unknown from the basis of something known: when you know the length of one side of a triangle and two of its angles, you can work out the length of the remaining two sides 119

model an idealized or simplified representation of reality that seeks to illuminate some of its characteristics 85

modernization a process of change based on *diffusion* and adoption. Apparently less developed societies adopt the characteristics of apparently more advanced – in economic terms – societies 133, 143, 165

morphology the science and study of form (*see geomorphology* and *landscape morphology*) 40, 41, 42

morphometric analysis the measurement and analysis of shape. A form of description using geometric, spatial co-ordinate systems when systematizing and classifying 88

multi-attribute region region based on a synthesis of a number of *geofactors* 56

nationalism both the feeling of belonging to a nation and the political ideology that is promoted to define a state as a territorial unit 31

natural landscape landscape little modified by human activity 46

natural regions division of the world based on morphology, climate and vegetation 154

necessary conditions mechanisms that need to be in place in order that events may happen. However, necessary conditions are not sufficient in themselves to explain beyond doubt the reasons why the events happened. See *realism* 129, 130

neighbourhood an area, mostly in an urban community, in which face-to-face relationships predominate. For example, a relatively small-scale, residential unit that has been planned to cater for local needs (school, kindergarten, small shops), that is closed to through traffic and that has an identity that is distinct from other neighbourhoods 170, 171, 175

network studies a term used in *spatial science* for studies of transport networks; in sociology and cultural geography it denotes studies of local and regional social relationships 84, 170

'new regional geography' regional studies established in Britain in the 1980s focusing on the persistence of regional diversity in the face of the homogenization of global capitalism 155

nomothetic law giving. Primarily concerned with regularly occurring phenomena or known processes rather than with the explanation of specific instances (*idiographic*) 10, 11, 13, 57, 69, 72, 73, 104

normal science in the Kuhnian model, the type of science undertaken when a discipline is dominated by a ruling *paradigm* 60

normative question a question about values; about what a thing ought to be rather than what it is 102

normative theory concerns what 'ought to be' rather than what is, was or will be (*positive theory*). *Spatial science models* were used to create a world that is 'rational'. Such models and the theory behind them have hence been labelled 'normative' 91, 116

objectivity external reality, things as they are independent of our mind 106, 107, 110

ontology the theory of being and the nature of existence. Empiricism and realism may be regarded as two ontological traditions 115, 117, 118, 125, 128, 132, 134

organism analogy something which seems to be similar to an organism, a 'whole' 39

otherness concept used to define contrast to mainstream cultures 167, 168

outsidedness as opposed to a person's *insidedness* in a place; an outsider may be a visiting tourist or a researcher 160, 194

palaeontology the science of the forms of life existing in former geologic periods as represented by fossils, animals and plants 40

paradigm the academically accepted understandings within a discipline of the discipline's field (aims and tasks) and scientific procedures (exemplary research methods) 58–62, 69, 72, 85, 86, 97, 98, 188, 190

partial-general reasoning the study of particular phenomena but with the ultimate aim of presenting general statements (as in the laws of physics) 4

partial-special reasoning the explanation of individual phenomena that are thought to be unique; *idiographic* reasoning 4

participant observation a qualitative research approach in which the researcher observes the subjects' everyday lives by participating in those lives. An attempt to acquire insider understanding 70

peneplain according to W. M. Davis, the flat plain created as the end result of the cycle of erosion 42

perceived environment   the subjective perception of the *phenomenal environment* that surrounds a human being. As presented by Olavi Granö, the perceived environment is converted (on the basis of an individual's beliefs, knowledge, experience and intentions) to a *cognized environment*, which forms the basis for action   121

perception   the subjective evaluation of the world around us   96

permanent revolution   in scientific theory, the idea that an active science should always fight against the establishment of ruling *paradigms*   61

phenomenology   a philosophy that tries to capture the universal and general structures of the world as it exists in our minds prior to any (*empirical*) scientific investigation   95, 121, 127, 158

physical classification   in Immanuel Kant's categorization, the grouping together of phenomena according to their position in time or place. Phenomena that follow each other in time are studied by history (*chronological science*). Phenomena grouped together in the same place are studied by geography (*chorographic science*)   24

physical geography   the study of the physical features of the earth, their causes and relations to each other   26, 41, 88, 187

physikalische Geographie   one of the three-fold divisions of the natural sciences proposed by Alexander von Humboldt   26

physiography   the description of nature. A term widely used for the geography of nature, wider in scope than physical geography   26, 41, 42

place   a portion of geographical space. Sometimes defined as 'territories of meaning' or 'a node of activities'   3, 54, 127, 145, 153, 155–163

placelessness   non-local, international general features created throughout the world by globalization, modernization and internationalization   137, 162, 163

plastic space   space whose perceived size or travel-time distances are constantly changing as a result of changes in technological progress and socioeconomic demands   75, 76

political geography   spatially informed political studies, including welfare geography and electoral geography   165

polymath   a person with knowledge and scholarly training in many disciplines   25

positionality   status or standing in relation to place or location. Positionality depends on what sort of a person you are: whether you are an insider or an outsider in a particular place   167

positive theory   descriptive theories that seek to account for what is, was or will be on the basis of empirical observation. In contrast to *normative theory* that is used to promote an 'ought-to-be', rational perspective on the world   91

positivism   a philosophical system developed by August Comte which claims that science provides the only valid form of knowledge and that sense perceptions of observable phenomena provide the only possible road to scientific knowledge. Directly opposed to *metaphysics*, and further developed in *logical positivism*   37, 98, 99, 100–106, 110, 113, 115, 124, 127, 189

possibilism   a concept first developed by Lucien Febvre in opposition to environmental determinism. Possibilism emphasizes the importance of choices in human activity rather than the limitations to choice   38, 45, 68, 70

postindustrialism   modern society in which a majority of the workforce is employed in tertiary and quartenary industries   133, 195

postmodernism   a philosophy of science that is related to *poststructuralism* but, more directly, that was established as an (empirical) reaction to modernist changes in societies. Postmodernism advocates pluralism and, in geography, the place-specific in contrast to the *placelessness* of modern societies   124, 132–138, 162, 167, 185

poststructuralism   a philosophy of science related to *postmodernism* but, more directly, based on a (philosophical) critique of modern knowledge (as, for instance, of the hyper-reality created by the modern mass media). It has fundamental scepticism towards any form of *structuralist* explanation, capitalist as well as Marxist. Methodologically, emphasizes the need for a critical understanding of texts and language and the crucial importance of *self-reflection*   112, 124, 132–138, 190

practical consciousness   knowledge we see as more or less self-evident, which we use in our daily lives and which we cannot necessarily explain in words. Connected to Giddens' *structuration* theory   126

preparadigm period   according to Kuhn, the first phase in the model of the development of science, in which there are different schools of thought clustered around different scientists   60, 69

probability model   a model based on the probability that a given cause will have given effects   81

problem formulation   the crucial start of any scientific investigation   66
process model   a model based on current trends and processes   68
professionalization   the process of making a job market open only to those who have special educational qualifications   34, 60
projection (in cartography)   a systematic two-dimensional transformation of the three dimensions of the spheric globe   21
pseudo-constraints   assumptions, taken-for-granted rules of society, that can and may be changed   109, 117
puzzle solving   a term used by Kuhn for achieving (anticipated) results within the framework of a ruling paradigm   60, 86

qualitative methods   methods used to examine the social world, in which the central focus is on the definitions, behaviour and intentions of the agents studied. Used particularly in *fieldwork* and in *case studies*, the methods employed vary widely: from empirical approaches (e.g. *grounded theory*) and *hermeneutic analysis* to approaches connected with *critical theory* and *poststructuralism*   70, 112
quantitative methods   statistical and mathematical methods used to analyse quantifiable data sets and to build and analyse *models* and *systems*   112
quantitative revolution   a widely used but rather misleading term for the transformation that occurred in the 1950s and 1960s in Anglo-American geography in particular. The quantitative revolution led to the establishment of the *spatial science school*   73, 76, 82, 84
quarternary period   the last of the major subdivisions of the geological scale of time, including *pleistocene* (1 000 000–10 000 years before present time with a number of glacial periods) and the recent epoch (10 000–present time)   41
quasi-determinism (*quasi* = resembling or seemingly)   a polemic concept that was used to argue against the Marxist view that individual actions are unconsciously (in reality, only *seemingly*) manipulated by the ruling economic forces   123

radical geography   the study of the radical changes that may occur in the way societies are organized. The objective is to foster changes in current society and therefore in its social geography. Some radical geographers advocate revolutionary change (*see also Marxist geography*)   93, 97
rational   agreeable to reason; rationalism implies that reason is the supreme source of knowledge   94
realism   a philosophy of science that uses abstraction to identify the *necessary* causal powers and the conditionalities of the structures that are realized under specific *contingent* conditions. Realism occupies a position between classical *empiricism* and *transcendental idealism*. While empiricism-positivism collapses the world into a singular plane of space-time events in its study of empirical regularities, realism seeks to recover the connections between different dimensional domains in order to identify the relations between structures, mechanisms and events   106, 124, 127–132, 138, 155, 158, 175, 190
realm   a domain or field, used by Sack in four connotations: *realms of nature, meaning, social relations* and *agency*, in the conceptualization of *relational space*   149, 150
region   a distinct segment of the earth that, in the traditional sense of the word, has developed a particular character through a long process of interaction between humanity and nature (*vertical connections*). Regions have also been defined on the basis of interactions between centres and hinterlands (*horizontal connections*)   39, 46, 73, 145, 153, 160
regional complex analysis   according to Haggett, a combination of the results of spatial and ecological analysis   12, 68
regional geography   the study of an area that takes into account an area's total composition or complexity. Alternatively, regional geography may focus on a particular aspect of an area in order to illuminate its special character   5, 14, 28, 45, 71, 73, 153
regionalization   the delimitation of regions   53, 56, 93
regional monograph   a geographical dissertation that analyses a particular region   47, 48, 59
regional planning   application of scientific knowledge to promote the development of a region   54, 55
regional science   a discipline concerned with the theoretical and quantitative analysis of regional economies and problems that combines economics, geography and planning   77
regional studies   general term including description and analysis of regions, regionalization, researcher's devotion to regional specialization   53

regional survey   observation and recording in the field, particularly connected to the approach recommended by Patrick Geddes   54

regional synthesis   an account of an area's special character with a focus on the relations between humanity and the natural environment   71

relational space   space and place are intrinsic parts of our being in the world – we relate to other people and the physical environment. Thus relational space is consciously or unconsciously embedded in our intentions and actions   142, 147–153, 173, 196

relative space   the location of, and distance between, different phenomena (*horizontal connections*) as the focus of geographical inquiry. Distance as measured in terms of transport costs, travel time and the mileage within a network, as well perceived distance, is given explanatory power   7, 75, 76, 83, 84, 146, 195

remote sensing   obtaining images by devices placed far off from objects, i.e. in sattelites etc.   180

resource   a source of human satisfaction. Natural resources are the things or forces in nature that, at the present level of technology and economic demand, can be exploited to serve human needs   177, 180

revolutionary phase   in Kuhn's *paradigm* model, the phase when one paradigm is overthrown and replaced by another   60

scientific law   according to one definition: a generalization of unrestricted range in time and space. Other definitions accommodate as scientific laws *empirical generalizations* that are valid only for a specific time and space   39, 57, 63, 64

self-reflection   in scientific work, the need for a researcher to reflect on his or her own role and impact within the research process   112, 134

semiotics   research into *signs* and meanings in order to uncover the underlying cultural codes that determine how people interpret the spoken word and other forms of human interaction   113, 123, 134

semiotic triad   a model of the relations between objects, signs and interpreters that is used to underline the relevance of semiotics in a research process   134

sense of place   an individual awareness of the 'spirit' or identity of a place. Sense of place may be related to physical properties (landscape, urban history), to practical layout and organization (traffic links, location of services), social relations (family, friends) and roots (personal history at this location)   95, 158–169

sequent occupance   stages in the development of a region, often brought about by a new immigrant culture   49

sign   a symbol (such as a word), an icon (such as map made to signify an object) or an index (a sign that relates to an object in the same way as smoke relates to fire). *Metaphor* is also a type of iconic sign   134, 135

simulation   through the use of a computer, modelling all the possible outcomes of a process while deploying all the rules that are considered to be decisive in that process's possible outcomes. Monte Carlo simulation models were used by Hägerstrand to solve the problems of the *spatial diffusion* of *innovations*   81

single-attribute region   region delimited on the basis of one geofactor, for example vegetation   56

singular   the only one of its kind   11, 188

singularity trap   the failure to recognize generalities – the results of general structural forces – in the study of a particular geographical area   143

social anarchism   as proposed by Peter Kropotkin, if the political structures of dominance and subordination were removed from society, better societies would develop through a process of co-operation. Centralized institutions limit economic progress and promote inequality. The ideas of social anarchism are opposed to both *social Darwinism* and to *state socialism* as promoted by Lenin   35, 36, 133

social Darwinism   originally proposed by Herbert Spencer. Human societies must struggle with their natural environments and with each other in order to survive. The fittest individuals survive best within an economic system founded on free enterprise   36

social exclusion   in geographic terms, the concentration of economic, political and cultural 'outsidedness' in certain localities, which excludes the people living in those localities from participating in mainstream society   167, 168, 169, 171, 172, 173, 174

social geography   in German usage the term often refers to an integrated human geography; in Anglo-American usage it refers to a branch of systematic geography that deals with the geography of social groups   11, 35, 50

social housing   housing provided for the public, today mainly for those who are not able to cope in the housing market   172, 173

social mobilization   the bottom-up involvement of citizens in the planning and development of their local communities   175

social physics   uses analogies from the physical world to describe or analyse aggregate human behaviour   80, 93

space   a central concept in geography, used in the form of *absolute, relative* and *relational space*   145

spatial analysis   the study of the relative localization and distribution of a significant phenomenon or group of phenomena   11, 177

spatial diffusion (see *diffusion*)   85

spatial exclusion   a social practice in which certain social groups are excluded from, or confined to, certain living quarters or activity spaces. This creates 'ghettos' of poor as well as 'ghettos' of rich people. Its most extreme form was the former apartheid system of South Africa   172

spatial factor   factor that have an influence depending on its relative location in space   5

spatial science   a definition of geography as the science of relations in space or as the science of spatial distributions. Connected to the so-called 'quantitative revolution' and to the concept of relative space   7, 76, 77–88, 96, 97, 104, 115, 116, 146, 153, 182, 187

special geography   a term first used by Bernhard Varenius to describe what is now called regional geography   14, 23

stochastic model   a mathematical-statistical model that describes, in probability terms, the outcomes from a series of trials. Sometimes a distinction is made between probabilistic models (in which the outcome of an individual trial is predicted as probable) and stochastic models (in which the development of a series of outcomes is modelled). Stochastic models are in contrast to *deterministic models*   81

structuralism   the study of the theory of the processes and structural forces that underlie and determine empirical events. We may distinguish between structure as construct type (as in structural linguistics), in which structures are seen as more or less permanent, and structure as process type (including structural Marxism), in which structures change through time   100, 117, 122–124, 188, 190, 194

structuration theory   as developed particularly by Giddens, a focus on the role of *agency* in interpreting, transforming and perpetuating structures. In contrast to *structuralism*, structures are not seen as determining empirical events but are regarded as the resources as well as the rules for actions   88, 124–127, 155, 190, 194

structural Marxism   (see *historical materialism*)   125

structure   the underlying and enduring driving forces within society, defined in a number of different ways in *structuralism, structuration theory* and *realism*   123, 125, 129, 173

subjectivity   internal reality, or reality as experienced within one's own mind   107, 110

sufficient conditions (see *necessary conditions*)   129

superstructure   in Marxism, the legal, political, religious and other ideological forms of domination. These forms of domination are guided by the 'infrastructure' or the economic base of production   106, 158

synthesis   a complex whole formed by the combination of its constituent elements   2, 11, 187, 191

systematic geography   branches of geography characterized by their theme of study, i.e. economic geography, population geography, geomorphology   3, 12, 14, 28, 71, 73, 153

systems analysis   the analysis of the structure and function of identifiable systems (*see* Box 3.5)   88–90, 93, 178

teleological explanation (*teleos* = purpose)   an understanding of events and phenomena in relation to their supposed underlying purposes in some predetermined design. Regarded as the opposite of *mechanical explanation*   27, 68, 69

temporal analysis   procedures for describing or explaining phenomena in relation to their development over time   146

territoriality   the geographical exercise of power. An individual or a group influences or establishes control over an area   151, 153, 158, 165, 166, 167, 168

theory   a coherent group of general propositions used to explain a class of phenomena (Newton's theory of gravity), or a proposed explanation whose status is still unproven. Theory and *hypothesis* are often used in the same way to mean an untested idea or opinion   65, 66, 91

thing language   the language of positivism that would supposedly unify science. Based on empirical observations and definitions, and formal logics   102, 107

time-space geography (also termed time-geography)   as developed by Hägerstrand and associates, time and space are regarded as limited *resources* on which individuals have to draw in order to realize particular projects. Time-space geography is considered to be central to Giddens' *structuration theory*, as time-space geography's graphical representations make it possible to appreciate the logics of structuration   81, 82, 93, 96, 127

topography   the description of places in some detail. In antiquity, topographical description meant general descriptions of places; today topography is understood as the relief features or surface configurations of an area   17, 19, 21, 22

topophilia   the affective ties a human being may develop with a place   160, 161

topophobia   contrary to topophilia, the repulsive feelings a place may create in a human being (for instance, 'landscapes of fear')   160, 161

total-general reasoning   the nature or essence of scientific objectives. This concept (as employed by Schmithüsen) is a bit vague, but may be regarded as the reasoning concerned with both *epistemology* and *ontology*   4

total-special reasoning   the analysis of the complex features of particular unities, such as landscapes and regions in geography   4, 146

transcendental   a philosophical stance based upon the doctrine that the principles of reality are to be discovered through the study of the processes of thought. The term is connected to certain trends in both *idealism* and *realism*   127, 128

uniform region   region characterized by overall sameness or homogeneity of man-made or natural features   154

universal theoretical law   scientific law with universal validity, fx. the physical law of gravitation   64

urban ecology   urban geography and sociology using analogies from ecology in analysing city growth and structures   50

urban morphology   study of forms of and within urban landscapes   71

Utopia   a place or state of ideal (unattainable) perfection, concept derived from the described society in Sir Thomas More's 'Utopia' (1516)   101

validity   logical soundness. To maintain that a scientific theory is *valid* implies that it is not proven to be true, but that it is sound and useful for the time being   64, 115

verification   'the confirmation of a hypothesis through a series of experiments. Verification or rejection have been regarded as the logical outcomes of a test of a scientific theory. Others would argue that absolute verification is, in most cases, impossible   63, 64, 100

vernacular (definition of a discipline)   a definition based on ordinary people's everyday or common understandings. For a school discipline, this will often mean in practical terms a field of learning it is given in the school curriculum by school politicians. This may contrast with an academic definition of the discipline   139, 141, 142

verstehen   implies both understanding and empathy; putting oneself in the other person's shoes   94, 104, 109, 121, 154

vertical connections/linkages   the dependency of people on their local, natural resources   74, 75, 143, 146, 154, 177, 195

Vienna Circle   established in 1907 and later led by Moritz Schlick. The circle developed *logical positivism*, which emphasized the need for the *verification* of any statement if it were to be acceptable scientifically. They opposed other approaches to knowledge, which they regarded as *metaphysical*   100, 101

welfare geography   the study of spatial variations in the quality of life and, as proposed by some practitioners, the setting out of proposals for social and spatial policies aimed at reducing inequalities in the quality of life   93, 96

Wissenschaft   a wide definition of scientific knowledge that includes all forms of methodological study (several of which are regarded as 'arts' in English-speaking countries)   104